The Drone Question and Answer Book
(The UAV Question & Answer Book Revised)

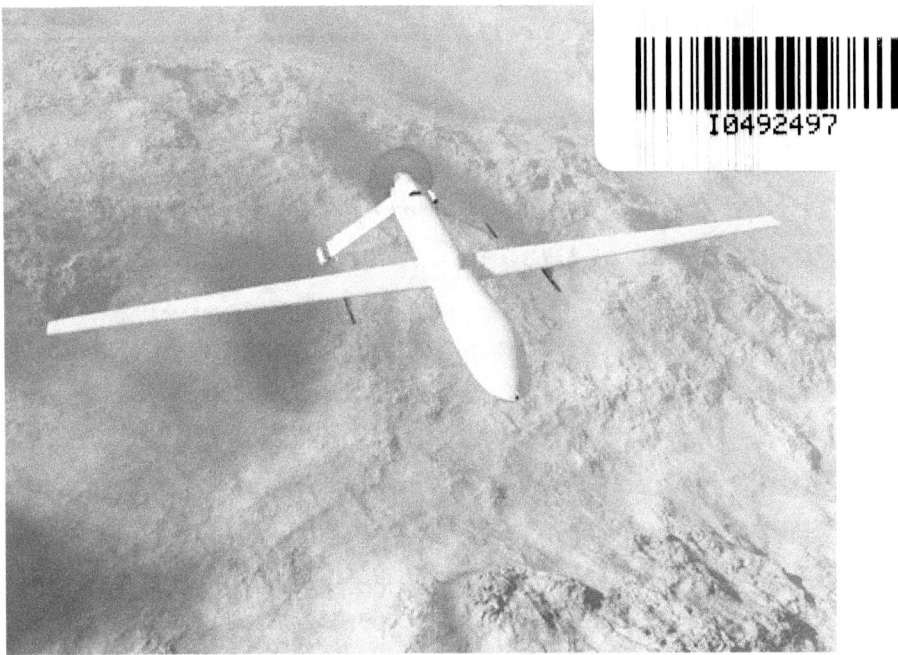

Predator in Flight

Author: Paul E. Love

Copyright 2017

Other books by the author:
- The Employer's Payroll Question & Answer Book (2017)
- Anatomy of a Google Sheets Project
- Cowboy Shooting Sports
- Virtual Robotics: Robotics on a Budget
- The Volunteer's Guide
- Rescue Me: Animals in Need

CONTENTS

Introduction

Are manned aircraft about to go the way of the Pony Express or silent movies? Has the last generation of manned fighter pilots already been born? Probably not, but it is clear that unmanned aerial vehicles (drones) are having an increasing impact both on the battlefield and on commercial air operations. Drones or UAVs (Unmanned Aerial Vehicles) have already become a vital part of the United States military arsenal and they are also finding new applications in civilian life every day.

Modern military drones are revising the the whole concept of aerial surveillance and close air support. Providing extensive coverage of the battlefield and delivering lethal payloads with surgical precision, drones are guided by satellite and can be operated by personnel sitting in a trailer halfway around the world. They have proven to be a major asset in locating and targeting terrorist activities in places like Iraq, Afghanistan and Serbia and in providing vital information to field commanders and troops on the ground.

Similarly, drones are showing up outside of the military doing jobs such as:
- Gathering data from inside a hurricane to aid in scientific research
- Performing border patrol duties for the U.S. Customs service
- Providing fire fighters with thermal imagery and mapping information on wildfires

- Taking part in search and rescue operations, aiding in disaster response, transporting cargo, performing crop dusting, and providing police with an "eye in the sky"

This book is concerned primarily with the use of drones in the United States and is intended to provide a quick overview of what drones are, how they're being used, how they operate and what impact they may have on the present and future.

Note: This book is dedicated to the men and women of the 32nd Tactical Reconnaissance Squadron who (many years ago) let a young Army soldier work alongside them.

Drones In General

How did drones come about?

The earliest use of an unmanned aerial vehicle in war may have happened in 1849 when the Austrians used unmanned balloons loaded with explosives in an attack on the the Italian city of Venice. Later, aerial torpedoes were used in World War One - these "flying bombs" used mechanical gyroscopes to maintain straight and level flight. They were launched in the direction of enemy forces and when they ran out of fuel they would fall out of the sky and explode on impact. Later drones used radio guidance systems to carry out reconnaissance missions under control of a human pilot. The invention of integrated circuits eventually led to the development of today's drones – unmanned aircraft that can fly themselves to a large extent and can also be operated by a ground controller.

What were some of the earliest drones?

Probably the first unmanned aerial vehicle was A.M. Low's "Aerial Target" in 1916. Lowe, an officer in the Royal Flying Corps in World War I, helped develop a remotely controlled aircraft that could be used as a guided missile. That was followed by a number of other remote-controlled aircraft, including the Hewitt-Sperry Automatic Airplane. Later, drones were used during World War II for target practice. Jet engine powered drones appeared following World War II but drones didn't really come into wide-spread usage in the military until the Vietnam War.

Were drones ever used in WWII to attack a target?
In July of 1944 a 4-drone attack was launched against a beached Japanese merchant ship in the Russell Islands. The attack resulted in two hits and two near-misses.

Where did the term "drone" come from?
The use of the word "drone" in connection with unmanned aerial vehicles may stem from the British development in the 1930s of an unmanned radio-controlled aerial target named the "Queen Bee".

Does the term "drone" only apply to vehicles that fly?
The word "drone" is also being used now to refer to both surface and underwater unmanned vehicles. Examples include US Navy underwater drones being developed for surveillance and possibly as unmanned attack vehicles. Police forces in the US are also beginning to refer to bomb-defusing and reconnaissance robots as "drones".

What was Operation "Red Wagon"?
After Gary Powers and the U-2 spy plane he was flying were shot down over the USSR in 1960, the U.S. started a classified drone development program called "Red Wagon". The program resulted in the development of drones (the Ryan reconnaissance drones) that flew over 3400 missions for the United States during the Vietnam war.

What is a "UAS"?

In many cases the term "Unmanned Aerial System" is used instead of "unmanned aerial vehicle" to emphasize that there is an entire system involved in drone operations, not just a "vehicle". In most cases the system includes one or more remotely-controlled aircraft, one or more control stations, and a communications and control link. For example, the RQ-7 Shadow UAS consists of four unmanned aircraft, two ground control stations (GCSes), one portable GCS and one Remote Video Terminal (RVT).

What is a "UCAV"?

An armed drone is often referred to as an unmanned combat air vehicle or UCAV.

What is an "MAV"?

Micro Air Vehicles or MAVs are proving to be effective tools for soldiers and police operating in the narrow confines of city buildings or crowded areas. MAVs (which were basically insect-sized) were limited by short flight times since they couldn't carry much fuel or heavy-duty batteries because of their size, and currently there aren't any true MAVs. DARPA (the Defense Advanced Research Projects Agency) is experimenting with Nano Air Vehicles with a wingspan of 7 ½ centimeters. The original MAVs included:

- XQ-138 – Developed at the University of Kansas, the XQ-138 could take off and land on its tail. It could fly at speeds up to 150mph by using actuators to counter wind gusts and keep it stable in flight.

- Cyberquad – Built by Cyber Technology, the electric-powered Cyberquad used four ducted fans that let it fly and hover like a helicopter. The Cyberquad was used primarily for short reconnaissance missions such as inspecting a damaged oil rig in the Timor Sea.

What does "sUAS" stand for?

sUAS stands for Small Unmanned Aircraft Systems. Advances in technology made it possible to pack the capabilities of older, larger drones into smaller and smaller airframes and the term SUAS refers to these smaller drone systems.

What does "STUAS" stand for?

STUAS stands for Small Tactical Unmanned Aircraft Systems and refers to micro-drones.

What does "RPA" stand for?

In the last few years the U.S. Air Force has begun referring to larger drones like the Predator, Reaper and Global Hawk as Remotely Piloted Aircraft (RPA) to emphasize that these systems are always controlled by a human operator. The term Remotely Piloted Vehicle (RPV) has also been used.

What's the difference between a cruise missile and a drone?

One basic difference is that the drone is reuseable and the cruise missile is not. In addition the drone is capable of executing a search pattern over a certain area and can be used for reconnaissance, surveillance and target illumination, as well as being able to attack and destroy a target.

What's the difference between a Vietnam era drone and current drones?

Unmanned aircraft like the ones employed by the U.S. in Vietnam were used for reconnaissance and basically just flew in a straight line or preset circles taking imagery until they ran out of fuel. Because of their relatively simple nature they began to be referred to as "drones" rather than UAVs. As new models were developed that were controlled by an operator the military began to use the term "RPV" (Remotely Piloted Vehicles). The general term "drone" began to be used to refer to all types of UAVs as unmanned aircraft began to combine remote operation with onboard piloting capabilities, ranging from simple flight plan programming to systems that allow the drone to determine its own flight path and maneuvers.

What's the difference between a drone and a R/C (radio controlled) aircraft?

Originally, to be considered a drone an aircraft's primary purpose needed to be military or commercial, not recreational. In general aircraft weighing less than 55 pounds aren't considered "vehicles" unless they have considerable autonomous operational capability, which in the past meant that normal R/C models weren't considered drones. However, with the explosion in small recreational drones, the difference between the two is mostly aesthetic. If it looks like a regular airplane it's an R/C aircraft – otherwise it's a drone.

Are military drones "flown" by a human pilot?

Yes and no. Most of the larger drones are capable of flying a pre-programmed flight plan with the human pilot taking control only if it's necessary to make some change in the route.

What are the basic types of drones?

There are two basic varieties of unmanned aerial vehicles – ones that have to be controlled from a remote location and ones that can be controlled manually or fly autonomously (on their own, without human guidance).

What is an "autonomous drone"?

Autonomous drones are capable of carrying out certain types of flight operations on their own, such as:

- Determining their own flight path based on their programming and on-board sensor readings
- Selecting the best maneuvers to perform a specific task
- Initiating communications with ground stations and other drones

Where will the biggest improvements in drones be made in the near future?

The current trend seems to be moving away from improvements in the technology that lets pilots fly the drone and toward greater autonomy for the drone itself. More sophisticated electronic systems are being developed to let the drone "think" for itself and,in the case of a military drone, make tactical decisions (short of launching an attack) on its own.

What are the different functional categories of drones?
- Reconnaissance - provide battlefield intelligence for the military
- Combat – armed drones that provide an attack capability in high-risk environments
- Target – provide ground and aerial gunnery targets
- Logistics – drones designed for cargo carrying
- Civilian – drones specifically designed for commercial use
- Research and Development – used to test new designs and new technologies

What types of military missions are assigned to drones?
Unmanned aerial vehicles fly reconnaissance or attack missions (or both). Drones can also provide laser targeting for attack helicopters.

Are there any anti-drone weapons?
Obviously smaller drones can, under the right circumstances, be brought down by a baseball bat. Larger military drones though require something a little more sophisticated. U.S. Air Force researchers have developed a couple of different anti-drone weapons. One approach uses high-energy bursts of microwaves to disrupt a drone's electronics and disable the aircraft (such as the "Counter UAV" program put together by Boeing, Air Bus and Space), while a second approach uses a vehicle-mounted laser (such as Boeing's High Energy Laser Mobile Demonstrator) to shoot down enemy drones. Both programs reflect a growing concern over enemy

forces using small unmanned aircraft to spy on or even attack U.S. Forces.

A third approach is to hack into the drone's computer system and there have been numerous examples of that in recent years, particularly with military drones. One major incident involved a U.S. RQ-170 Sentinel that went down in Iran in 2011. The Iranian government announced that their cyberwarfare unit had gained control of the drone and commanded it to land in Iran. Other examples of "drone hacking" include reports that the NSA has hacked into the video feed of Israeli and Syrian military drones and that the terrorist group Islamic Jihad managed for several years to hack into the video feed of Israeli surveillance drones.

What are some other anti-drone weapons?
Other anti-drone weapons include:
- DroneShield introduced their "drone gun" in 2016 – another version of a signal-blocker. DroneShield's gun weighs 13 pounds and has an effective range of a little over one mile. Once a drone is engaged the drone gun transmits a signal commanding the drone to turn around or land.
- Openworks' SkyWall fires four projectiles at an incoming drone and comes in shoulder-fired, stand-mounted and turret-mounted versions. The four projectiles include a ballistic round, a basic net to tangle up the drone, a net with a parachute to try to bring the drone to the ground, and a "last-ditch" signal-jamming projectile.

- Weapons such as a signal-jamming gun used by China's police have also been developed for use against smaller (generally non-military) drones.

How long can a drone stay airborne?
Longer than a manned aircraft in many cases (assuming the manned aircraft doesn't refuel in-flight). In 2006 a solar-powered Zephyr drone stayed aloft for 54 hours and some current military drones can remain airborne for flights lasting close to 40 hours. The majority of commercial drones can stay in the air for up to 30 minutes while a few (mainly winged drones) can stay up as long as 45 minutes.

Can drones refuel in flight?
Not yet, but soon. On July 1st, 2010 DARPA (Defense Advanced Research Projects Agency) gave a $33 million dollar contract to Northrop Grumman to develop an aerial refueling system using two Global Hawks. According to initial press releases the system will also allow drones to refuel at much higher altitudes than conventional aircraft, which normally refuel somewhere between 20,000 and 35,000 feet.

Update #1: In 2015 the Navy's X-47B drone was successfully refueled in flight from a K-707 tanker.

Update#2: The Navy plans to have their MQ-25 Stingray refueling drone in operation on aircraft carriers by the early 2020's. The Stingray will help extend the range of carrier-based fighters and it also offers a much smaller radar signature than conventional manned tanker aircraft.

Are there drones that can "refuel" themselves?
Projects are underway to create drones that never have to land. The theory is that drones equipped with magnetometers should be able to recognize the magnetic field surrounding an electric power line. Once it detects the presence of a power line the drone could head towards it, "land" on the power line and recharge its batteries, making it capable of remaining airborne almost indefinitely.

Are all drones controlled from the ground?
Normally drones are handled by a controller on the ground. However, experimental flights were done several years ago to test a drone Command and Control Interface (droneCCI) that would allow a single manned fighter to control multiple drones while airborne. The drones are semi-autonomous, meaning that they can fly straight and level on their own and the pilot of the manned fighter only has to give them orders for specific actions such as attacking a specific target or executing a search pattern.

Are there drones that can accompany manned fighter jets?
The U.S. Air Force is currently testing two UAVs from Kratos Defense and Security Solutions that are capable of flying in tandem with F-16 and F-35 fighters. Both UAVs use AI (Artificial Intelligence) to copy the movement of nearby airplanes, which allows them to fly alongside a manned aircraft

without the need for a ground controller. The two Kratos drones are:

- The UTAP-22 Mako, which can fly at speeds up to 700 mph and deliver a mix of weapons.
- The XQ-222 Valkyrie, which has a range of 4,000 nautical miles and a top speed of 650 mph.

How are drone flight plans programmed?

Drones like the Reaper or Global Hawk are usually set up to follow a set of "waypoints" fed into the drone's GPS (Global Positioning System) receiver. Waypoints are a set of coordinates (longitude, latitude, altitude and possibly time) that identify a particular point in space (or space and time).

Has a military drone ever gone out of control in a civilian area?

Yes, at least on one occasion. In August 2010 Navy operators lost control of an MQ-8B Fire Scout. The operators finally regained control but not before the drone entered restricted airspace around Washington, D.C. The incident was apparently the result of a software problem that caused the Fire Scout to ignore its pre-programmed flight procedures.

What's the smallest drone?

AeroVironment is testing its "Nano Hummingbird" drone, which is designed to look and fly like a hummingbird. The Nano Hummingbird has a wingspan of just 6 ½ inches (16 cm), weighs just 2/3 of an ounce (19 grams), and can climb, descend, move sideways, fly forward and backward and clockwise and counterclockwise. In spite of its tiny size, the Nano Hummingbird contains a motor, batteries, a video camera, and a communications system so it can be flown by remote control.

What's the largest drone?

Currently AeroVironment's Global Observer is the largest drone in operation. Global Observer measures 70 feet in length, with a wingspan of 175 feet. It can operate at 55,000 to 65,000 feet and stay aloft for up to 7 days while carrying a payload of up to 400 pounds (180 kg). Military mission capabilities include acting as a satellite-type communications system, allowing different teams to communicate over long distances. Possible civilian uses include weather monitoring, emergency communications link in case of natural disaster, and coastal surveillance.

What is the largest military drone?

Probably the Navy's Triton UAV (the successor to the Global Hawk). Developed by Northrop Grumman, Triton has a wingspan of 130 feet (roughly the same as a Boeing 757) and can cruise at altitudes of up to 50,000 feet.

Can drones program themselves?

Not exactly, but Stanford University created a drone helicopter that can learn to fly complex stunts by "watching" another aircraft perform those maneuvers. The helicopter gets its flight plan by recording everything that a human operator does while flying another aircraft. Then the helicopter's GPS system and avionics communicate with a ground-based computer to work out a flight plan.

How do the international laws of war impact drone operations?

Laws of war such as the Geneva Conventions place a burden of combatants to limit collateral damage through proper target identification. In order to be able to assign responsibility for the actions of a drone a human has to be "in the loop" - completely autonomous drones are not an option. In addition the person "in the loop" should be a member of the military who understands and accepts his or her role as a combatant.

Military Use of Drones

When did the U.S. first consider using unmanned aircraft in combat?

The idea has been around for a long time. Consider the following quote from General Henry Arnold (who commanded the Army Air Forces during WWII):

"We have just won a war with a lot of heroes flying around in planes. The next war may be fought by airplanes with no men in them at all."

General Arnold may have been a little premature but that prediction is coming closer to reality all the time.

Where did the U.S. first use drones in combat?

During the Vietnam war RPVs (remotely piloted vehicles) equipped with video cameras flew reconnaissance missions over North Vietnam to spot troop concentrations, supply routes, and anti-aircraft sites. RPVs were also used to do electronic jamming and to drop propaganda leaflets. The RPV was launched from a C-130 and piloted by two officers aboard the C-130. Once the mission was complete the drone was maneuvered close to a CH-53 helicopter, the RPV's parachutes were deployed, and the CH-53 snagged one of the parachutes and carried the drone back to its base.

Who first used drones to attack battlefield targets?

During the Yom Kippur War in 1973 Syrian missile batteries in Lebanon damaged or destroyed so many Israeli fighter jets that Israel turned to unmanned aircraft research for an alternative to manned fighters. That research led to their development of the first modern drone. The effectiveness of Israeli drones allowed them to attack and destroy virtually all Syrian surface-to-air missile sites in southern Lebanon at the start of the 1982 Lebanon War. The success of the Israeli drones caught the attention of other nations and spurred the development of unmanned aerial vehicles by the military in other countries.

How fast is the use of military drones increasing?

U.S. Military use of drones has increased so rapidly in recent years that the military is having a hard time keeping up. Flight hours over Afghanistan and Iraq more than tripled between 2006 and 2009, but only about a third of the requests by ground commanders for drone assistance are being filled due to a shortage of aircraft and controllers.

What drones do the Air Force and Army fly?

The Army primarily flies RQ-7 Shadow and RQ-11B Raven drones and the MQ-1C Gray Eagle (an updated version of the MQ-1 Predator). The Air Force uses both the smaller drones like the Raven and ScanEagle and the larger drones like the MQ-1 Predator, MQ-9 Reaper and the RQ-4 Global Hawk.

How is the Navy using drones?

Remember the stories about the Navy training dolphins to carry and attach magnetic mines to enemy ships? Well, the dolphins may be off the hook. The Navy has plans to make use of groups of underwater drones which can use a variety of sensors to identify and attack enemy ships, gather oceanographic data, and conduct reconnaissance missions. Small underwater drones can relay data to Navy ships almost in real time by using satellite integrated telemetry, providing the Navy with a major tactical advantage.

What is the "Tier system"? ** Deprecated

The U.S. Military uses a "Tier" system to categorize the different drones each service uses. For example the Air Force tier system breaks down this way:
- Tier N/A – small/micro drones (currently handled by BATMAV drone)
- Tier I – Low altitude, long endurance flights (currently handled by the Gnat 750)
- Tier II – Medium altitude, long endurance (currently consists of MQ-1 Predator and MQ-9 Reaper)
- Tier II+ - High altitude, long endurance (RQ-4 Global Hawk)
- Tier III – High altitude, long endurance low-observable (RQ-170 Sentinel)

In comparison, the Army's Tier system includes the RQ-11 Raven (Tier I), RQ-7 Shadow (Tier II), and the RQ-5A/MQ-5A/B Hunter (Tier III).

What is the "Group System" for classifying US military drones?

The tier system (which classified drones primarily by function) has been replaced by the "group" system. Military UAVs are divided into the following groups:\
- Group 1 – weight (0 to 20 pounds), speed (up to 100 mph, operating altitude (less than 1200

feet above ground level), members (WASP, RQ-11 Raven)
- Group 2 – weight (21 to 55 pounds), speed (less than 250 mph), operating altitude (less than 3500 AGL), members (ScanEagle)
- Group 3 – weight (less than 1320 pounds, speed (less than 250 mph), operating altitude (less thn Flight Level 180), members (RQ-21 Blackjack, RQ-7B Shadow)
- Group 4 – weight (greater than 1320 pounds), speed (any), operating altitude (less than Flight level 180), members (MQ-1A/B Predator, MQ-1C Gray Eagle, MQ-8B Fire Scout)
- Group 5 – weight (greater than 1320 pounds), speed (any), operating altitude (greater than Flight Level 180), members (RQ-4 Global Hawk, MQ-9 Reaper, MQ-4C Triton)

How much did the U.S. use drones in Iraq?
During the war in Iraq, drone missions accounted for 80% of all flights over Iraq.

How were drones used during Operation Iraqi Freedom?
Drones were used extensively right from the beginning:
- Prior to the start of Iraqi Freedom Hunter drones were flown near Iraq's border with Kuwait to gather reconnaissance on Iraqi movements, AA sites and other information.
- Hunter drones were also used to locate where enemy forces were storing ammunition and weapons.

- Army Shadow drones were used to perform perimeter guard duty around U.S. Camps.
- drones helped keep the Republican Guards from making it to Baghdad to reinforce the troops there by pin-pointing the location of the Guards units and by guiding attacks against them.
- drones also provided protection for U.S. convoys and ground troops, warning them of enemy activity in the area and possible ambushes or booby traps.
- Real-time drone day and night surveillance allowed Coalition forces to detect enemy activity and mount an attack within minutes.
- Predator and Reaper missile attacks and laser targeting helped to disrupt and destroy Iraqi communications and control.

Are drones taking over from manned aircraft in the US military?

Yes, to some extent. In 2009 the U.S. Air Force trained more drone pilots than fighter and bomber pilots combined. And in 2010 the U.S. bought more unmanned aircraft than manned aircraft. However, current planning in the Air Force envisions manned fighters leading swarms of drones into battle. DARPA released a concept video in 2015 showing swarms of small, expendable drones scouting ahead and relaying information back to a (manned) stealth fighter. The fighter in turn would relay that information to a cargo plane which could launch missiles and missile-carrying drones to carry out attacks.

Does the Air Force Academy have a drone training program?

In 2009 the Air Force Academy started a drone training program, with cadets flying Viking 300 drones over Fort Carson to get hands-on experience flying a drone. Since then the program has been greatly expanded and in 2015 the Air Force announced that it would open up drone training to enlisted personnel.

How many drones does the U.S. Military have?

As of 2009 the U.S. Military had over 7,000 drones, ranging from hand-launched Ravens to jet-powered Global Hawks. By mid-2016 that had increased to roughly 9,350 UAVs, including approximately 340 MQ-1 Predators, MQ-9 Reapers and RQ-4 Global Hawks and almost 8,000 hand-launched drones. And that's just the beginning - within the next 20 years the Air Force expects to have entire squadrons of unmanned fighters, bombers and tankers, along with thousands of bug-sized drones.

What types of missions do drones fly in Afghanistan?

drones like the Global Hawk and Raven are used for surveillance and reconnaissance. Others like the Predator and Reaper carry out actual combat missions, either firing missiles themselves or illuminating targets with laser pointers for helicopter pilots.

What are the main advantages of using drones?

Aside from the obvious advantage of not putting aircrews at risk, drones are able to stay airborne for 24 hours or more carrying out surveillance over a given area without being detected. That allows analysts to study possible enemy activity over a period of time, make sure of their targets and coordinate an attack to take out those targets while avoiding civilian casualties. And even in cases where decisions have to be made in a hurry there are a number of people in the loop which should result in better decisions and fewer mistakes. There is also a major benefit for troops on the ground who get to see images relayed from drones overhead, showing them what's waiting around the next bend in the road or past the next building. And in situations where troops run into trouble armed droneS orbiting the area may be able to move in quickly to take out snipers, mortar teams and other threats.

What are the main disadvantages of using drones?

Military drones currently in use are very susceptible both to advanced enemy anti-aircraft weapons and to interruptions in their communications linkage (either from natural causes or from enemy action). As newer drones come into service they will be able to fly higher and faster and may include electronic counter-measures, which should make them less vulnerable to air defense systems. A possible solution to the communications problem is a manned aircraft that would serve as an airborne controller, handling one or more drones remotely and removing the need for long range communications links. For long-range communications, enhanced electronic security measures will help to prevent an enemy from hacking into the satellite signal. An additional problem with current drones is the slight time lag in communication between the drone and its operators

– a factor that should also improve as the technology involved improves. Finally, the lack of physical cues and limited peripheral vision can make a drone more difficult to fly than a manned aircraft - a combination of experience, improved technology and training may eventually help to minimize that problem.

How does the weather affect drones?
Wind can definitely be a factor with small drones. Larger versions like the Predator are also temperamental about the weather – rain for example can soak into the carbon-fiber skin of a Predator, adding weight to an airframe that is already on a carefully balanced weight to fuel ratio. Certain weather conditions can also degrade the communications link between a drone and its controllers.

What happens when the communications link to a drone is broken?
When a "lost link" occurs, drones like the Predator and Reaper simply start to circle on a preset course and wait for the link to be re-established.

Which military drones carry weapons?
So far the only armed drones are the MQ-1 Predator and the MQ-9 Reaper. Originally a reconnaissance drone, the Predator was modified to carry two AGM-114 Hellfire missiles or other munitions. The Reaper, the first UAV designed for high-altitude long-endurance reconnaissance, can carry a variety of weapons including Hellfire II air-to-ground missiles, AIM-9 Sidewinder missiles and GBU-12 Paveway II laser-guided bombs. There may also be a third armed UAV in the inventory before long. Duke

Robotics, based in Florida, has developed a drone that is equipped with machine guns.

Are Air Force drone fliers "pilots" or "operators"?
The officers who fly armed drones are officially referred to as "pilots" - however some pilots who fly manned aircraft believe the title should be "operators" since drone pilots are not physically in the aircraft. As the Air Force transitions more and more to drones though, it's becoming a moot point – the people who fly the drones do the same job and are in fact "pilots" just as if they were sitting in a cockpit.

Do drone pilots receive flight pay?
In late 2009 the Air Force got the go-ahead to begin giving drone pilots flight pay. It's not clear yet whether sensor operators will be able to draw flight pay, but both pilots and sensor operators will be able to wear wings for their uniforms, similar to those worn by manned pilots and aircrew members. The wings, flight pay and other Incentives are part of the Air Force's planned shift toward a largely unmanned fleet by 2047.

In addition to flight pay, the Air Force announced in 2016 that experienced drone pilots who agree to re-enlist for five years will be also be eligible for Aviator Retention Pay bonuses of up to $35,000 per year.

Is "drone pilot" a specific Air Force occupational code?
Drone pilots now receive an "18X" Air Force Specialty Code – it is now a "rated" career field and comes

with a six-year active duty service commitment. Sensor operators will receive the "1U" Specialty Code, requiring a 3-year active duty commitment.

Is there resistance within the U.S. Air Force to the use of drones?

In a word, Yes. There seems to have been a fair amount of resistance, especially at first, from the upper levels of the Air Force. However there was also resistance to the use of cruise missiles rather than strike aircraft during the Cold War and just as cruise missiles became a standard part of the Air Force, drones are quickly becoming an indispensable part of Air Force operations.

What is an MQ-1 Predator?

The MQ-1 Predator (made by General Atomics) is a "Tier II" medium altitude, long endurance (or MALE) drone used primarily by the U.S. Air Force and the CIA. The Predator carries cameras and other sensors for reconnaissance missions and has also been modified to carry and fire two AGM-114 Hellfire anti-tank missiles or other munitions. Powered by a gasoline engine and propeller driven, the MQ-1 has a range of up to 400 nautical miles and a loiter time over target of about 14 hours. Predators have also seen civilian use for border patrols and for scientific studies.

How long can a Predator drone stay airborne compared to an F-16?

An F-16 can stay over a target area for about an hour before it has to land or refuel from an airborne tanker. In contrast a Predator can linger over a target area for up to 24 hours and use about a tenth the amount of fuel consumed by the F-16.

How big is a Predator?
A Predator is twenty-seven feet long with a 49 foot wingspan and weighs about 1200 pounds without fuel or missiles.

What does a Predator UAS consist of?
A Predator Unmanned Aircraft System or UAS consists of four Predators, a ground control station (GCS) that houses the pilots and sensor operators, a satellite link communication suite, and operational personnel.

What is a "MAC"?
Most Predator and Reaper missions in Iraq or Afghanistan are actually flown by personnel in the U.S. working in a van known as a Multi-Aircraft Control system or MAC. Two officer pilots and four enlisted sensor operators in the MAC can fly up to four missions at a time. The sensor operators consoles are almost identical to the pilots', allowing the enlisted men to take control of the drone. The sensor operators fly the drone inside an airspace pre-defined by the pilot - generally the pilots only take over during an emergency or to fire weapons.

How many people are in a Predator squadron?
A typical Predator squadron consists of 350 to 400 people, including 27 pilots.

What exactly is an MQ-9 Reaper?
The Reaper, also referred to as the Predator B, is a medium-to-high altitude drone whose primary

objective is to act as a hunter/killer against targets of opportunity. Its secondary mission is to act as a surveillance and reconnaissance platform, using various sensors to relay real-time data to field commanders and intelligence specialists. The Reaper has a complete sensor suite including an infrared sensor, a color/monochrome daylight TV and an image-intensified TV. Reapers can carry a variety of weapons including the AGM-114 Hellfire missiles, GBU-12 Paveway laser-guided bombs, the AIM-9 Sidewinder air-to-air missile, and the GBU-38 JDAM (Joint Direct Attack Munition). In addition the MQ-9 has a laser rangefinder/designator giving it the capability to designate targets for laser-guided munitions.

How big is a Reaper?
An MQ-9 Reaper is thirty-six feet long with a 66 foot wingspan and weighs about 4900 pounds without fuel, missiles or bombs. The Reaper also flies at about 300 miles per hour (compared to about 100 miles per hour for the Predator).

What does a Reaper UAS consist of?
A typical Reaper UAS consists of multiple drones, a ground control station, communications equipment, spare parts, operations and maintenance personnel. Like the Predator, the Reaper is crewed by a pilot and a sensor operator.

How do Predators and Reapers compare to fighters like the F-15 or F-16 costwise?
The MQ-1 Predator costs about 7 ½ million dollars and the Reaper runs about 13 ½ million, compared to 70 or 80 million for an F-16.

Are all Predators and Reapers used in Afghanistan and Iraq flown by operators in that country?
Predator and Reaper missions are actually flown by operators stationed in the U.S. at places like Creech Air Force Base near Las Vegas. However there is a lag of about 1 ½ seconds in the satellite signal from overseas which is too long a delay for a pilot trying to take off or land, so drones are still launched and landed by personnel in Iraq or Afghanistan.

How many Predator and Reaper missions are flown every day in Iraq and Afghanistan?
As of 2009, an average of about 30 unmanned combat air patrols were being flown every day.

What kind of sensors does a Predator or Reaper have?
Predator and Reapers contain a full suite of sensors: electro-optical, infrared, video, and laser/target markers.

What type of armament does a Reaper carry?
A weapons loadout for a Reaper normally consists of four Hellfire missiles and two 500 pound GBU-12 laser-guided bombs.

How long can a Predator stay airborne?
Predators often stay in the air for up to 24 hours (with a shift change in operators).

How many Predators have crashed to date?
As of the fall of 2009, approximately one third of the 200 Predators delivered to the U.S. Air Force had

crashed, either because of aircraft malfunction or operator error. However, better documentation and more intense training programs now in operation appear to be helping to reduce that percentage.

What is a Global Hawk?
The RQ-4 Global Hawk is a high-altitude, long endurance surveillance and reconnaissance drone which can provide area coverage in all weather, day or night. The Global Hawk UAS consists of the aircraft, its sensors, a ground station, data link, and trained personnel. Different versions of the RQ-4 are being developed and fielded in "capability blocks". Block 10 is currently operational and carries imagery intelligence (IMINT) sensors.
Note: Block 20 will include a larger aircraft with increased payload capacity and improved IMINT. Block 30 will have both the improved IMINT sensors and integrated signals intelligence (SIGINT) sensors to achieve simultaneous, multi-INT coverage. Block 40 will comprise a separate set of aircraft employing an advanced radar sensor to provide improved IMINT and surface/air moving target information.

How much does a Predator, Reaper or Global Hawk cost?
Military drones like the Predator, Reaper and Global Hawk generally cost tens of millions of dollars. According to one report a Predator costs about 4 to 5 million dollars, a Reaper about $11 million (half as much as an F-16) and a first-generation Global Hawk runs about $35 million. As of a couple of years ago, the cost of a second-generation RQ4-B Global Hawk had increased to around 56 million dollars. NOTE: cost estimates for military drones can vary widely

depending on whether the figures are for the drone only or for an entire UAS, including ground control and communications equipment.

Does the U.S. Army fly Predators?
As of 2009 the Army has units using the MQ-1C Gray Eagle, an enhanced version of the MQ-1B Predator.

What other drones does the Army use?
Two of the primary drones the Army uses are the RQ-11B Raven and the RQ-7 Shadow.

What is an RQ-7 Shadow?
The RQ-7 Shadow drone is used by the U.S. Army and Marine Corps to carry out reconnaissance missions. It's launched from a trailer-mounted pneumatic catapult and recovered using arresting gear similar to that on an aircraft carrier. The RQ-7 is a little over 11 feet in length with a wingspan of 14 feet and weighs about 185 pounds when it's empty. It has a maximum speed of 135 miles per hour, a range of 68 miles and a maximum altitude of 15,000 feet.

What does an RQ-7 Shadow UAS consist of?
A Shadow UAS includes four drones, two fixed ground control stations (GCSes), one portable GCS, one launcher, two Ground Data Terminals (GDTs), one portable GDT and one Remote Video Terminal.

How many people are needed to operate a Shadow UAS?
Currently it takes 22 people to fully man a Shadow drone system.

Are other countries using drones?
More than 40 countries currently have drones in their aircraft inventory and more are on the way. For example, China has at least 25 models of drones currently in service including the Combat Eagle stealth UCAV and the Thunderbolt UCAV. One day in the not-too-distant future, U.S. drones may be engaging enemy drones in unmanned aerial dogfights.

What are some of the drones used by other countries?
Foreign made UAVs incude:
- The Heron, made by Israeli Aerospace Industries (IAI). Weighing over 1,000 kilos with a wingspan of over 16 meters, the Heron can fly for up to 52 hours at a height of 10,000 meters (35,000 feet) - roughly the same height as a commercial airliner. The US, Canada, India, Turkey, Australia, and Morocco have all bought Herons for tactical and reconnaissance use.
- The German military uses the LUNA, made by the German company EMT Penzberg. LUNA is a reconnaissance drone, which the Bundeswehr has relied on for several thousand hours of flight time in Afghanistan and Kosovo since 2000. It's considerably cheaper than the Heron, but its range is only around 100 kilometers. Pakistan and Saudi Arabia have also bought LUNAs.
- The Black Hornet, a 1 inch by 4 inch micro drone manufactured by the Norwegian

company Prox Dynamics, has been used by the British army since 2013. The Black Hornet can hover for up to 25 minutes on one charge of its battery, is equipped with night vision and an infrared sensor, and the digital data link to its terminal has a range of up to a mile (1.6 kilometers). It's been used by British soldiers to look around corners and over walls in Afghanistan since it went into service in 2013. Once it's launched the Black Hornet can be controlled from a small handheld terminal which also displays the transmitted images from its three cameras.

Could drones be used in terrorist attacks in the United States?

At some point drones may become a major concern in regards to terrorist attacks. For example small or micro drones could be used to track people and provide information to allow terrorists to set up an ambush – or a small drone filled with a deadly toxin could be flown into a stadium or other crowded area and the toxin released by remote control.

How accurate are drone attacks?

There have been questions raised about the accuracy and targeting of drone attacks. Israeli drones armed with missiles killed 48 Palestinian civilians in the Gaza Strip in 2009; those casualties were blamed on the failure of operators to distinguish civilians from combatants. Another problem with drone strikes is the weapons involved – for example the MQ-1 Predator is normally armed with 100 lb Hellfire missiles which were designed to take out tanks and to attack hardened bunkers. Smaller weapons such as the Raytheon Griffin and the U.S. Navy's Spike missile are being developed to offer a weapon load capable of a more surgical attack. Along with increased accuracy, missiles like the Griffin also allow the Predator A to carry six missiles rather than two of the much heavier Hellfires.

Update (October 2016):
Recent data released by the U.S. Government has fueled arguments over the accuracy of drone strikes. Opponents of drone use say the statistics released show that drone strikes kill more non-combatant civilians than piloted air strikes. However the figures don't take into account the improvement in accuracy of drone strikes over time or the difference between manned aircraft attacks on specific targets versus drone attacks on targets-of-opportunity, which are generally spontaneous and carry a higher risk of non-combatant casualties.

What kind of engine do most drones use?
Smaller drones are generally powered by small rotary or two-stroke piston engines while some of the large drones like the Reaper are jet-powered.

Do military drones crash more often than manned aircraft?
Actually, despite the fact that drones fly more hours than manned fighters or bombers, their safety record is about the same. drones don't have redundant systems like manned aircraft (since crew safety isn't involved) and that contributes to the number of drone crashes, as do pilot error and control signal interruptions. In addition because of the demand for drones the software itself has some weaknesses –

for example, one Predator was lost because the pilot inadvertantly managed to erase the onboard RAM (Random Access Memory) making the drone impossible to control.

Do military drones fly according to a program or are they actually piloted by a human?

It depends on the particular drone. Some, like the Global Hawk, carry out their missions according to programmed instructions – ground controllers can transmit updated programming to the drone but they can't pilot it directly. Other systems fly on autopilot most of the time (except for takeoff and landing), but can be controlled directly by a human pilot when necessary.

What are ground control stations like?

Military control stations are normally mobile and can be in a building, a vehicle, or another aircraft. Most of them consist of two workstations with one used by the pilot and the other by a sensor specialist (although either workstation can be used to pilot the drone in case the pilot's console has a problem). In some cases there is also a third station for a "mission intelligence specialist" whose main job is to coordinate with friendly ground forces. The station's electronics package normally allows the drone to be controlled either by direct radio communication or by satellite link when the drone is beyond range of the line-of-sight data link.

What does a Predator control station look like?

The pilot and sensor operator sit in front of an array of monitors. There are two primary monitors – one shows the video feed from the Predator's cameras overlaid with a display of the horizon, the Predator's altitude and other vital information while the other monitor displays a graphic image of the drone overlaid on a satellite map of the area.

How does a drone operator make a Predator or Reaper change directions?
The "pilot" draws a line on the screen of his monitor and the drone alters its pre-programmed flight path, following the changes drawn on the monitor.

Do you have to be an experienced pilot to be allowed to pilot drones?
The Army and Marines rely primarily on enlisted personnel who have never flown, but they are only dealing with small, unarmed drones. Until recently the Air Force only allowed pilots who had flown real aircraft (fighters, helicopters, heavy transports, etc.) to pilot the larger drones, particularly the Predator and Reaper armed drones. However, due to the pressure to add more pilots the Air Force began a new program in 2009 to train officers with no prior time in a cockpit as drone pilots. The program involves six months of screening and basic flight training followed by a nine-week initial qualification course including forty hours in a simulator and at least nine actual flights. So far results have indicated that non-pilots do about as well as experienced fliers.

How receptive are Air Force pilots to being assigned to fly drones?
There seems to be a mixed reaction, with pilots who have been flying helicopters or "heavies" (such as

transports or tankers) being more receptive than fighter pilots to transitioning to drones.

Author's Note: Years ago when I was in the Army I spent about 18 months as liaison to an Air Force squadron flying Army requested missions. The aircrews I worked with flew Phantoms and were bright and dedicated officers. However, my feeling is that neither the pilots or the backseaters (Phantoms had a 2-man crew) would have been happy being taken out of the cockpit and put in a drone control station. I suspect that's still true of fighter pilots today – they have a dangerous job (even if they're not flying in combat) and I think part of what attracts them is testing themselves physically and mentally every time they go up – that and the "need for speed".

What are the basic duties of an Army drone operator?

Army personnel assigned to the 15W MOS or Unmanned Aerial Vehicle Operator Military Occupational Specialty supervise or operate UAVs such as the Shadow reconnaissance drone (and its supporting systems) to gather intelligence for mission planning. Specific duties include:

- Plans, prepares and helps carry out air reconnaissance missions.
- Operates mission sensors to locate and identify targets.
- Deploys and redeploys the Tactical Unmanned Aerial Vehicle (TUAV) ground and air system.
- Operates and performs operator level maintenance on communications equipment,

power sources, light and heavy wheeled vehicle and some crane operations.
- Launches and recovers the drone.
- Performs pre-flight, in-flight, and post-flight checks and procedures.
- Directs emplacement of the ground control station and launch and recovery systems.
- Supervises and assists in air frame repair.
- Coordinates replacement of parts and other necessary supplies.

What kind of training is required to fly an Army Raven drone?

After 10 weeks of Basic Combat Training operators must complete 23 weeks of advanced training, learning skills such as:
- Preparing charts and maps
- Preparing intelligence reports
- Performing simulated surveillance and reconnaissance missions
- Analyzing aerial imagery
- Using advanced computer systems

Overall, according to actual trainees, the hardest part of the course may be mastering the technique of hand launching drones.

Where is the Army's drone training center?

Fort Huachuca in Arizona is home to the world's largest drone training center, training approximately 2,000 Army, Marines, and foreign military students each year. Over 350 employees of General Dynamics teach at "Black Tower", training students on the Hunter, Shadow, Gray Eagle and Warrior Alpha systems. Students at the Fort Huachuca facility spend at least half their time flying simulated drones – in some cases soldiers complete 100 hours of simulation compared to just 10 hours of flying an actual drone. Fort Huachuca is not only the largest drone training center in the world it's also the only place in the world where unmanned aircraft have priority over manned aircraft.

Where is the Air Force's drone training center?
Currently the main training center for the U.S. Air Force is at Creech Air Force Base in Nevada (Note: until 2005 Creech AFB was an auxiliary airfield for Nellis AFB). Pilots at Creech Air Base also fly most of the Predator and Reaper missions in Iraq and Afghanistan (although launches and landings – which require quick response - are handled on site by drone operators in Iraq and Afganistan, due to the slight lag time in communications between there and the U.S.).

Do Air Force drone pilots have to meet the same standards as manned aircraft pilots?
At present the Air Force may relax some physiological requirements, but drone pilots must meet the same requirements for eyesight as manned aircraft pilots and the same academic standards, including passing the Test of Basic Aviation Skills and the Air Force Officer Qualification Test.

What is Air Force drone operator training like?
First of all, except for some small short-range reconnaissance drones, drone pilots have to be qualified officer pilots. Sensor operators however are normally enlisted personnel and have the ability to take control of the drone when necessary. Flying

a drone requires the same basic skills as flying any aircraft. However the emphasis for drone pilots and sensor operators may be more on information processing than on actual flying. Semi-autonomous drones can handle a good deal of the mechanical part of flying on their own – the primary job of the operators is to assimilate the information being passed to them and make appropriate decisions based on that data.

A good deal of the training involves learning how to use the drone's sensors to best advantage. For example, during mid-day in warm weather, a Predator's infrared imaging can tell you if a vacant parking lot has had vehicles parked there recently. The spaces where the vehicles were parked will be cooler than the rest of the lot since they were shaded from the direct sun.

What characteristics make a good military drone operator?

- Proficiency at flying the drone
- The ability to concentrate for considerable periods of time when very little is happening – to stay alert and watch carefully for any items of significance.
- Talent at developing tactics "on the fly" - the ability not only to pick out obvious targets, but to do "pattern analysis", to study the enemy's habits, where they store supplies, what routes they like to take, where they generally come from and where they go.
- The ability to work closely with others – especially since drone controllers usually work

as a two-man team, with one person doing the piloting and the other analyzing the imagery from the drone.

What is the hardest part of piloting an armed drone?

The most difficult thing about flying a Predator or Reaper is the same as with a jet fighter – deploying the weapons. Trying to coordinate with troops on the ground who want you to attack a target that has friendly troops and/or civilians within 50 feet from it can be extremely challenging.

What advantage does a drone pilot have over a fighter pilot?

Aside from personal safety, a drone pilot may spend up to three years flying missions over a certain area compared to a few months for a fighter pilot. That allows the drone pilots to become much more familiar with the territory they're observing – which makes it easier for them to spot changes in behavior, increased traffic in a certain area, or other things that may indicate terrorist activity. Familiarity with the people and daily life in the operational area can also help differentiate between real insurgents and innocent civilians.

What are some disadvantages of being a drone pilot or sensor operator?

One disadvantage is the amount of information drone operators have to deal with. Not only are there readouts from multiple sensors, there are the flight controls, the communications link, and the need to coordinate with ground troops or other personnel.

Also, with the current shortage of drone pilots, workload presents an additional problem. Constantly changing schedules in particular can upset a pilot or sensor operator's internal clock, causing mental stress and fatigue. An additional concern for drone crews is the feeling of impotence in cases where they can see events happening that are putting troops or civilians in danger and they aren't in a position to do anything about it. There is also some psychological stress in being involved in battlefield operations for hours and then going home to normal family life – that constant adjustment and the lack of fellow soldiers to talk with during off-duty hours can be difficult.

What is the "multiple-mission model"?
Some experts feel that the solution to the shortage of drone pilots is to advance the technology that allows a single pilot to control multiple missions. The multiple-mission model is a framework that may lead to a pilot being able to conduct a dozen or more missions simultaneously.

How similar is flying a drone mission to flying a jet fighter?
Very similar according to pilots who have flown both types of aircraft. For example every Predator mission starts with mission planning, including weather, route and intelligence briefings, the same as for a manned flight. Once the mission planning is complete the drone crew moves to the ground control station and goes through a pre-flight checklist just as they would for a manned fighter, except that the checklist is done over the radio with the drone ground crew. The control station itself is

arranged like a cockpit, with the pilot sitting in the left-hand seat and a sensor officer in the right-hand seat. The ground station controls include a stick and rudder, but with multiple computer screens taking the place of most of the other buttons, switches and warning lights that you would find in a normal fighter cockpit. Piloting the drone also is also very similar to flying a fighter according to the air crew, with the main exception being that there aren't the same physical cues as in a manned aircraft – no sound or vibration and limited peripheral vision (a fact which can actually make the drone somewhat harder to fly and land). Also, drone support and maintenance procedures are about the same as for a jet fighter and require a ground crew of about 50 people who are responsible for the weapons, electronics, radio and video gear, airframe and engine.

What is involved in mission planning?
Mission planning is the creation of a flight plan that takes into account:
- Terrain
- Weather
- Aircraft performance capability and configuration
- Weapon and/or cargo payload
- Fuel requirements
- Assessment of the route based on any known enemy threats
- Checking of the flight route and timing versus other aircraft in the area to avoid conflicts

What is the AFMSS?

The Air Force Mission Support System provides automated mission-planning support. As far as drones are concerned the system has been used to flight plan Global Hawk missions.

What are the advantages to a flight crew of flying a drone rather than a manned aircraft?
The primary advantage of a drone is the lack of outside distractions – not only the operators' knowledge that they're not personally in danger but also the lack of gut-wrenching high-g maneuvers or other physical stress on the crew. Also, many drones can stay airborne for long periods of time and in those situations a new crew can take over after a certain number of hours, ensuring that crew members stay alert and ready to respond quickly to any situation.

How much does it cost to train a fighter pilot versus a drone pilot?
According to an article in the Air Force Times it takes more than a year and over 2 ½ million dollars to train a fighter pilot, compared to 20 weeks and $135,000 to train a drone pilot.

Can a military drone maneuver like a jet fighter?
It depends somewhat on the type of drone. The MQ-1 Predator is powered by a small piston engine (115 hp) and attempting hard turns at high speed can send it out of control. Later models like the MQ-9 Reaper have a much more powerful turbofan engine (about 950 hp) and are capable of more high-speed maneuvers. However no current drone can match

the ability of a fighter like the F-16 or F-22 to accelerate, turn, climb, or dive.

What exactly happens when a drone fires a missile?

When the pilot fires a missile the sensor operator uses a laser designator to guide the missile to its target. If the target is moving the sensor operator has to keep it in the crosshairs of the laser designator until the missile gets there, which can be 30 seconds or more.

Who gives permission for a drone to fire on a target?

If the drone is flying a combat surveillance mission the personnel in the ground control station normally make the decision to fire. If the drone is performing a CAS (Close Air Support) mission in support of troops engaged with the enemy, then the JTAC (Joint Terminal Attack Controller) on the ground in Iraq, Afghanistan or wherever has to give clearance to fire.

Are there drones that can take off and land on an aircraft carrier?

There are smaller drones that can be employed on a carrier, but in February 2011 the U.S. Navy conducted the first flight of the X-47B, an F-18 sized drone that will be carrier-capable. The X-47B is designed for stealth and is meant to serve as a carrier-launched robotic strike aircraft. Since the X-47B is jet powered it can fly much faster than prop-driven drones like the Predator and Reaper and should be able to carry out a range of missions.

Currently the X-47B is slated for carrier trials sometime in 2013.

What is the UXV Combatant?

BAE Systems is working on designing what amounts to a drone carrier for the British Royal Navy. The warship is expected to displace approximately 8,000 tons and feature two decks for launching drones, VTOL (Vertical Take Off and Landing) aircraft and helicopters. The UXV Combatant reportedly will also be able to launch unmanned submarines and medium-range missiles and is expected to enter service with the Royal Navy sometime after 2020.

Can an enemy gain information from our drones?

There have been several instances where insurgents have gained intelligence of their own from our drones. In one case, when one U.S. combat division would end its tour of duty, it would ground all its drones until the new division moved in and took over. The insurgents figured out that pattern and took advantage of the "down time" to carry out operations while they knew they weren't being watched.

Are there any all-drone fighter squadrons?

Yes, the New York Air National Guard 174th Fighter Wing began transitioning from F-16 fighters to MQ-9 Reapers in 2008 and became the first all-drone attack squadron.

Why were baby wipes important in Iraq?

Baby wipes were used to clean the sand and other grit and grime off of the Army Hunter and Shadow drones in order to keep them flying.

Has an enemy unit ever surrendered to a drone?

Oddly enough, yes. During the Gulf War Iraqi Army forces surrendered to the drones of the USS Wisconsin.

Has a drone ever been in a dogfight with an enemy fighter?

A drone has engaged a jet fighter on at least one occasion. In 2002 a U.S. RQ-1 Predator exchanged missile fire with an Iraqi MIG-25, with the MIG's missile destroying the Predator.

Are there drone test pilots?

The U.S. Military just recently appointed its first drone test pilot, Captain Nicolas Helms, stationed at the USAF test pilot school at Edwards Air Force Base in California. His job will be to "push the envelope" and establish the performance limits of various drones.

What is the Air Force's "UAS flight plan"?

In 2009 the U.S. Air Force approved the "Air Force Unmanned Aircraft Systems Flight Plan (2009-2047)", outlining a strategy for integrating UAS operations with all Air Force core functions and joint force priorities.

What is an ARSS?

The Autonomous Rotorcraft Sniper System or ARSS is one of the U.S. Army's answers to the problem of fighting guerrillas who can attack and then hide among the local citizenry. The ARSS is a small unmanned helicopter (a Vigilante 502 drone) equipped with a .338 sniper rifle. An autopilot handles the business of flying the drone while a human operator on a remote monitor fires the rifle. To keep the cost down a good deal of the hardware in the ARSS is off-the-shelf, including an Xbox 360 video game controller that operators use to control the turret. Other weapons are also being considered for use with the ARSS including the M-249, an AA-12 shotgun, or non-lethal weapons.

What are "Micro drones"?
Miniature drones are an area getting increasing attention. Micro vehicles range from drones that can be carried by an individual infantryman to man-portable models that can be carried along with an infantry unit and launched like an anti-aircraft missile.

What is the "LMAMS"?
The LMAMS (Lethal Miniature Aerial Munition System) is a man carried missile with a number of drone characteristics, including loitering time and a hand-held ground station.

I've read that the last manned fighter pilots have already been born ... is that true?
Never say never ... but very unlikely. There will always be a need for humans in the cockpit. When line-of-sight and satellite communications fail, you have to have the option of having human "eyes in the sky".

What will the next generation of military drones be like?

New types of drones are being introduced almost every week, so it's hard to define a "next generation". However there are some interesting models currently under development:

- The U.S. Navy is working on a bat-winged bomber (designated the X-47B) designed to take off from an aircraft carrier. The X-47B is expected to be part of the new generation of drones that can evade radar and fly much faster than propeller-driven Predators and Reapers. The X-47B is also an example of a shift in emphasis from the production of short-range tactical strike drones only to the development of longer-range strategic bombers.
- Aurora demonstrated their "Skate" drone at the Farnborough International Airshow in July 2010. The Skate is a little larger than a pizza box and is made from a Styrofoam-like material. It has two electrically powered tiltrotor propellers, onboard sensors, and a downlink capability that allows It to connect to a Windows or Linux PC. The tiltrotors let the Skate switch quickly between vertical and horizontal flight, somewhat like the V-22 Osprey that the U.S. Marines use.
- AeroVironment's Global Observer high altitude, long endurance (HALE) drone made it's first test flight under hydrogen power in January 2011. Global Observer has a 175 foot wingspan and can use solar power and hydrogen cells to drive four electric motors,

allowing it to stay aloft for up to a week at altitudes of 55,000 to 65,000 feet. It's intended to provide satellite-type coverage of a given area at a fraction of the cost.

- Electric drones kept aloft by laser power-beaming technology may provide almost unlimited flight time.
- The Defense Advanced Research Projects Agency (DARPA) is supposed to be working on the design of a drone that will be able to stay aloft without any maintenance for up to five years.
- Some experts even predict that in the not-too-distant future drones capable of vertical take-offs and landings will detect enemy activity, land and release small unmanned robots armed with various types of weaponry. The robots (possibly guided from the drone once it's airborne again) will carry out the actual attack on the opposing forces, taking the place of human troops on the ground.

Civilian Use of Drones

Are there restrictions on where and how you can fly a drone in the United States?
In a word, yes – there are a number of restrictions for both recreational and commercial drones.
Restrictions for recreational drone operators include:
- Pilot requirements – None.
- Aircraft requirements – Drones that weigh over 0.55 pounds must be registered with the Federal Aviation Administration by filling out an application at www.federaldroneadministration.com.
- Location – Must stay at least 5 miles from any airport unless you've received prior permission from the airport and air traffic control.
- Operation – Must keep your drone in sight, must weigh under 55 pounds and must always yield the right-of-way to any manned aircraft.

Restrictions for commercial drone operators include:
- Pilot requirements – You must be at least 16 years old, have met the Transportation Security Administration (TSA) requirements, and have a Remote Pilot Certificate.
- Aircraft requirements – Must be registered if it weighs over 0.55 pounds, must weigh less than 55 pounds, and must undergo pre-flight checks to ensure that the drone is safe to fly.
- Location – Must stay in Class G airspace (uncontrolled airspace that extends from the ground to the base of Class E airspace).
- Operation – Must keep the UAV in sight, fly only during daylight, fly at speeds at or below

100 mph and altitudes under 400 feet, must not fly over people or from a moving vehicle, and must always yield to any manned aircraft.

How do I get a remote pilot's certification?

If you are a non-pilot applicant (you don't already have a pilot's license of some type) then you need to comply with the following steps to obtain a remote pilot certificate:

1) Locate a center for taking the FAA remote pilot knowledge test through CATS or PSI/LaserGrade Computer Testing (the FAA maintains a complete testing center list online).
2) Review the Remote Pilot–Small Unmanned Aircraft Systems Airman Certification Standards to understand the aeronautical knowledge standards for the remote pilot certificate with a small UAS rating.
3) Take and pass the FAA Unmanned Aircraft General—Small (UAG) Knowledge Test and obtain a knowledge test report. You must obtain at least a passing score (70 percent) on the knowledge test, which consists of 60 multiple-choice questions.
4) Apply for the remote pilot certificate through IACRA (the FAA's Integrated Airman Certification and Rating Application) or in writing using the paper application. Applying online generally means a significantly shorter processing time and the ability to obtain a temporary certificate.
5) Register in the FAA's IACRA system as an "applicant" and confirm the name on the FAA

knowledge test report is the same as the name used for registering in IACRA. Submit an online application for a remote pilot certificate using IACRA (no visit to an FAA-authorized individual is necessary because the knowledge testing center will verify your identity). The FAA will send you instructions via email to access a printable temporary certificate online and your FAA-issued permanent certificate will arrive in the mail.

Is there a demand for drone operators?

The job market for drone operators is still small, but it's growing right along with the boom in commercial drone use. A 2013 report released by the Association for Unmanned Vehicle Systems International, projected that more than 100,000 new jobs in unmanned aircraft will be added by 2025.

What are some non-military jobs where drones are being used?

- The U.S. Customs and Border Protection agency currently uses MQ-9 Predator B drones to patrol the border between the United States and Mexico. Infrared cameras mounted in the Predator's nose transmit images by microwave link back to Fort Huachuca in Arizona. When an observer there spots something suspicious the information is relayed to a Border Patrol Station, which sends a Blackhawk helicopter to investigate. If necessary the Predator can then illuminate people or vehicles on the ground with a laser designator to guide the Blackhawk directly to the target.

- Scientific research that involves obtaining data from remote or hazardous locations, such as flying into the center of a hurricane to get meteorological data. For example, in 2005 the National Oceanographic & Atmospheric Administration (NOAA) flew an Aerosonde drone into the heart of tropical storm Ophelia, gathering data in winds of 175 mph.
- Cargo carrying – helicopter or blimp drones can haul various types of payloads at a low cost.
- Fighting wildfires – Modified Predator Bs from NASA have been used to map forest fires and provide thermal imagery that can be passed on to firefighters on the ground. The information gained from the imagery can help locate the areas of maximum danger to firefighters and to analyze the spread of the fire.
- Monitoring oil and gas pipelines – above-ground pipelines can be carried out by digital cameras mounted on drones.
- Crop dusting – drones have been used in Japan for crop dusting.
- Celebrity watching – Supposedly a celebrity-photo agency is developing a camera-equipped paparazzi drone to track and photograph celebrities.
- Search and Rescue – The U.S. Coast Guard has tested the Bell Eagle Eye Tr-916 as a ship-based search and rescue platform.
- Mapping – Some testing has been done with an eye to using very small, light drones with video cameras to do urban mapping somewhat like Google Earth and Google Maps.

- Environmental monitoring – In October 2010 a Meridian drone at the University of Kansas made its first flights over the Arctic, gathering data to monitor the movement of polar ice.
- Disaster response – Drones are projected to become an integral part of response to natural disasters in the future – unmanned aircraft with infrared (heat) sensors could be life-savers, spotting victims trapped in damaged buildings or homes.

Have drones actually been used in civilian search and rescue operations?

Drones were used successfully in the aftermath of the 2008 hurricanes that hit Louisiana and Texas. Photos from before and after the storm highlighted areas of damage and helped to direct rescue efforts. In addition, an RQ-4 Global Hawk was sent from Beale AFB to Haiti in January 2010 to assist in the humanitarian aid operations there following a 7.0 magnitude earthquake. Drones have also been used successfully outside the U.S. on many occasions, including locating and saving migrants attempting to cross the Mediterranean and locating survivors after major earthquakes in Nepal and Ecuador.

Update:
The use of drones in emergency and search-and-rescue operations in the U.S. was limited for years by the FAA, which banned drone operations in the National Airspace without specific permission from the FAA. Then in June of 2016 the FAA announced its "Part 107" regulations. Part 107 provides operational rules permitting businesses to fly drones

in the U.S. Providing the drones weigh less than 55 pounds and meet a set of FAA requirements. The drone pilot must also have passed an unmanned aircraft operator test. There is still a daylight only restriction on search-and-rescue operations, but Part 107 also offers a procedure that can be used to acquire a waiver (good for 4 years) that would allow night-time SAR flights.

Have drones been used for security purposes?
India used drones during the 2010 Commonwealth Games to provide surveillance for the Games Village and the different venues. The primary purpose for the use of drones was to prevent a possible airborne attack by a terrorist group using paragliders.

Has a non-military drone ever been used to kill someone?
Yes, on at least one occasion, after an hours-long stand-off with a sniper who shot five Dallas police officers. Rather than risking the lives of more officers, the Dallas police used a bomb-defusing drone to place C4 plastic explosive near where the shooter had barricaded himself, then had the drone detonate the explosive, killing the gunman.

Won't civilian drones be a danger to other aircraft?
Currently the FAA restricts operation of drones to certain specific areas to avoid the possibility of collisions with planes or helicopters. However systems are under development which will use a combination of video cameras and sophisticated image processing to give drones the equivalent of a

human pilot's "see and avoid" ability to search for incoming aircraft.

Note: The FAA's transition from radar-based air traffic control to a satellite-based system may also make drone flights in civilian airspace less of a concern since under the satellite system (known as ADS-B) aircraft will continuously broadcast their position.

Update: The Automatic Dependent Surveillance-Broadcast or ADS-B system (which is part of the FAA's NextGen plan) is already being implemented and will be a requirement for most aircraft and airports by 2020. ADS-B will broadcast aircraft information such as speed, location, and route to air traffic controllers and other participating aircraft through the use of satellite radio signals and ground stations.

Have there been many incidents with drones being flown illegally?

The FAA says the number of reports from pilots, law enforcement and others of "potential encounters" with drones has been rising sharply, and warned that operating "around airplanes, helicopters and airports is dangerous and illegal." There are now more than 100 such reports a month, the FAA said.

What are the current FAA restrictions and safety guidelines for hobbyists flying a drone in U.S. Airspace?

Here are the FAA regulations for recreational drones as of December 2017:

- Fly no higher than 400 feet and remain below any surrounding obstacles when possible.

- Keep your sUAS in eyesight at all times, and use an observer to assist if needed.
- Remain well clear of and do not interfere with manned aircraft operations, and you must see and avoid other aircraft and obstacles at all times.
- Do not intentionally fly over unprotected persons or moving vehicles, and remain at least 25 feet away from individuals and vulnerable property.
- Contact the airport and control tower before flying within five miles of an airport or heliport.
- Do not fly in adverse weather conditions such as in high winds or reduced visibility.
- Do not fly under the influence of alcohol or drugs.
- Ensure the operating environment is safe and that the operator is competent and proficient in the operation of the sUAS.
- Do not fly near or over sensitive infrastructure or property such as power stations, water treatment facilities, correctional facilities, heavily traveled roadways, government facilities, etc.
- Check and follow all local laws and ordinances before flying over private property.
- Do not conduct surveillance or photograph persons in areas where there is an expectation of privacy without the individual's permission.
- Follow any community-based safety guidelines, as developed by organizations such as the Academy of Model Aeronautics (AMA).

Do I have to register my drone that I fly as a hobby?

Like almost everything connected with drones, the answer to this question could change at any time. Right now there's no requirement to register a "hobbyist or recreational" drone. However, anyone who flies a drone under FAA regulations Part 101 (commonly known as the Special Rule for Model Aircraft) could soon be required to register with the FAA. Congress has passed the National Defense Authorization Act, including language that would restore the drone-registration requirement for model aircraft operators.

Should recreational drones be registered?

In 2015 the FAA ruled that recreational drones must be registered – that ruling was challenged in court and eventually overturned. But people heavily involved in the drone industry have since expressed the opinion that registration should be required. Their position is that there are various UAV groups working to expand what the FAA allows commercial drone operators to do, including seeking broader authority to fly over people and to carry out unmanned delivery services. That will require new air traffic control abilities and a high level of coordination and having large numbers of unmarked recreational drones in the air complicates such efforts. With no way to assess responsibility the FAA will be very reluctant to give drone owners more freedom.

What is a "COA"?

The Federal Aviation Administration has issued Certificates of Authorization (COAs) in certain cases to allow the use of drones in civilian airspace. The number of COAs issued has been a trickle though compared to the demand.

Does weather pose a significant problem for commercial drones?
Weather conditions could pose a problem for some current commercial models. Those drones that only have electro-optical sensors could be "blinded" by heavy cloud formations – military models with synthetic aperature radar (SAR) can see through cloud formations and weather. Onboard weather sensors and de-icing gear are also important to make sure the weather doesn't pose a threat for commercial drones.

What kind of education do you need to be a civilian drone pilot?
A four year college education is helpful, but with or without a college degree you will probably need a private pilot license, a commercial pilot license and a few hundred flight hours as the pilot in command.

Is there anywhere I can see what's involved in getting a private pilot license?
There are several good books available on getting a private pilot license:
- "The Complete Private Pilot" by Robert E. Gardner
- "Your Pilot's License" by Jerry Eichenberger
- "The Pilot's Manual: Ground School: All the Aeronautical Knowledge Required to Pass the FAA Exams and Operate as a Private and

Commercial Pilot" by Aviation Theory Centre Ltd.

Or you can go to:

http://www.free-online-private-pilot-ground-school.com

for a look at the basics of flying an airplane, from navigation to flight instruments to airplane aerodynamics.

How much can a pilot with drone experience earn as a civilian drone pilot in the U.S.?

Currently experienced drone pilots can start at salaries as high as $100,000 a year. Salaries of $75,000 to $85,000 are possible with very limited drone experience if you have a degree in a field like engineering or telecommunications. As more drone pilots return to civilian life from the military, future salaries may depend on how quickly the civilian market grows (which will depend heavily on what changes are made in FAA regulations and how quickly they are made).

Are there passenger-carrying drones?

Not yet, but the Chinese debuted a passenger-carrying drone (called the Ehang 184) at the 2016 Consumer Electronics Show. There has also been some talk about building military drones that could transport troops over short distances.

What is the Parrot AR.Drone?

An early entry in the civilian drone market, the Parrot AR.Drone is a small four-bladed helicopter controlled via iPhone over a wireless network. The Parrot features two cameras and can potentially fly at altitudes up to 160 feet, beaming back video form the cameras. You can actually buy a Parrot drone online at several locations including amazon.com.

What is the "SenseFly Swinglet CAM"?
A small RC drone – basically a flying wing with stabilizers, a rear propeller, and a small on-board camera. You simply set up your GPS coordinates for the area where you want to get some aerial imagery, shake the Swinglet to start it up and then let it go.

What is the "Air-Mule drone"?
Urban Aeronautics has developed a drone called the "Air-Mule", designed to assist in medical evacuation. The Air-Mule will be able to carry medical equipment and fly into hard-to-reach places utilizing it's ability to take off and land vertically, similar to a helicopter.

Where can I learn to fly a drone?
- drone Flight School (www.droneflightschool.org) offers non-automated drone/UAS flight training, including classroom and simulator time as well as actual flight time with a sUAS. Material covered includes procedures for aileron, engine or elevator failure, use of flaps, crosswind landing techniques, proper glide slope approach and proper use of the throttle.
- Both the University of North Dakota and Kansas State University offer four year degree programs in drone pilot training and Unmanned Aircraft Systems.
- Embry-Riddle Aeronautical University has also created a new minor in Unmanned Aircraft Systems at their Daytona Beach, Florida

campus. Courses include Unmanned Aircraft Systems, Unmanned Aircraft Systems Operations and Cross-Country Data Entry, Operational Aspects of Unmanned Aircraft, UAS Robotics, and Unmanned Sensing Systems.

- As a first step you could try the 2nd R/C Flight School (http://www.2ndrcflightschool.com) which teaches you how to fly radio-controlled model planes or enroll at any certified flight school that can help you get your private pilot's license.

What if I want to make a drone of my own?
Try this webpage:
http://hacknmod.com/hack/make-a-drone-spyplane-using-the-arduino/
The tutorial shows you how to build your own drone using a handful of servos, a battery pack, a GPS unit and the Arduino programmer.

Traffic Control for Commercial Drones

Are there plans to create a traffic control system for commercial drones similar to the Air Traffic Controller system for manned aircraft?
Yes there are plans being developed to control the ever-growing volume of commercial drone traffic in the U.S. Estimates of commercial drone activity by 2020 range as high as 7 million UAVs and some system needs to be in place by then to help keep all those drones from crashing into each other.

Who is working on putting together a UAV traffic control system?
NASA, the FAA and a number of partners in industry are working together to find a solution to the problem. There have already been tests of a prototype UTM (UAS Traffic Management) system.

How would a drone traffic control system work?
Unlike the system for manned aircraft, traffic control for drones isn't expected to involve human controllers giving verbal instructions to drone pilots. Instead UAV pilots will depend on a largely autonomous electronic system which would allow them to input and receive flight information and various notifications.

What specific types of services could be provided to drone pilots?
Such a UTM system could:

- Notify drone pilots and crews of potentially hazardous weather conditions, such as high winds (which have a much greater effect on drones due to their light weight).
- Currently drones aren't allowed to fly beyond their operator's sight. A UTM might allow drones to fly further away from the operator, with a combination of satellites and cellular networks providing tracking information.
- Warnings could be issued to operators flying too close to a sensitive areas such as power plants and airfields or approaching areas where an emergency situation is in progress.

How soon would the system be operational?
Current plans call for the FAA to implement a traffic control system by 2025.

Would the system have any effect on the traffic control operations for manned aircraft?
Not in the near future. However, NASA is also conducting research into how to integrate drones into the controlled airspace occupied by crewed aircraft.

Future Uses of Drones

What are some possible future uses of drones?

Here are a few possibilities:

- Amazon has done some research on using drones to recharge electric powered cars while they're on the road – driving. The "refueling" drone would fly from a charging station to the vehicle requesting a "fill-up", identify the proper car and dock with it. Then the drone's batteries would feed power to the car and, once the recharge was complete, the drone would return to its base station, recharge itself and wait for the next call.
- Aerial dancing? Recently a swarm of Parrot drones took part in a UK talent competition as part of a dance act.
- Back in the 1950s *Flying* magazine carried ads for aircars which could transport us by ground or air and predicted that there could be one in every garage before the year 2000. That didn't work out but personal transport by drone may be a real possibility, although probably not in the very near future. The Chinese have debuted the Ehang 184, a drone which can transport one or more persons and companies such as Uber are investing heavily in research on passenger-carrying UAVs. The main problem isn't the technology, it's regulation – how to manage the traffic and who would be responsible in case of an accident.

- Wild fires can cause tremendous damage and threaten the lives of people living in the area and firefighters working to control the fire. At some point in the future drones may be available that can quickly reach the area where a wildfire has broken out. Then the drones can fly a pattern around the area and drop flammable canisters that will ignite a controlled burn, forming a firebreak and keeping the wildfire contained.
- Facebook announced recently that they are looking at utilizing drones to extend the Internet. Drones may be able to provide a wi-fi signal in remote areas where internet access has been spotty or unavailable up to now.

Drone Photo Gallery

Drone Ground Control Station

MQ-1 Predator in flight

MQ-9 Reaper

RQ-4/MQ-4 Global Hawk

RQ-11 Raven (hand-launched)

MQ-8 Fire Scout

USAF Drone Pilot's Wings

DJI Phantom 4 commercial drone

Parrot Drone

RC Models and Drones for Hobbyists

[This section has been left in as reference – since this was written hundreds of recreational drones have appeared and are sold almost everywhere.]

Note: Learning to fly radio-controlled model airplanes is frequently mentioned as a good way to introduce the basic techniques involved in flying drones.

What types of engines do RC aircraft use?
RC aircraft (with the exception of gliders and other unpowered models) use electric motors, internal combustion engines, or jet engines.

What is a "glow plug"?
Glow plugs are used in model airplane engines and are similar to spark plugs. A battery is used to heat the glow plug initially, then fuel is fed into the engine and the glow plug ignites the fuel just like the spark plugs in a automobile engine.

Why are RC internal combustion engines referred to as "nitro" engines?
The standard fuel for model airplane engines consists primarily of methanol with a certain percentage of nitro-methane mixed in. That mixture is suspended in an oil base, commonly castor oil or a synthetic oil of some type. The term "nitro" comes from the volatile nitro-methane part of the fuel.

What types of RC aircraft are there?

Basic types of RC aircraft include:

- Trainers – Trainers are designed to help you learn how to fly. They normally have the wing on top for maximum stability and are powered by electric motor or by glow plug engines.
- Sport planes – Sport airplanes make up a large part of the RC airplane market. They can be any size or shape and are better at performing aerobatic maneuvers than trainers.
- Warplanes – Warbirds are a popular type of RC aircraft and generally are built for higher performance than trainers or sport models.
- Aerobatic models – These models are built specifically for performing aerobatics and are generally low or mid wing planes with powerful motors.
- Float planes – Float planes are designed to land on water, but require a good deal of skill to fly.
- Helicopters – Single and dual rotor helicopters are a very popular variety of RC aircraft and tend to be fairly easy to fly.

Vintage airplanes, jets (powered by ducted fans or miniature gas turbine engines), gliders, and blimps are also available.

How hard is it to learn to fly an RC airplane?

Most RC enthusiasts will tell you that it's harder than learning to drive a car. Flight simulators can be a good way to get started without having to worry about crashing an expensive RC model.

What are "control surfaces"?

The control surfaces on a model airplane are the parts of the plane that control its movements – the rudder, elevator, ailerons, etc.

What are "servos"?

The control surfaces on a model plane are actually moved by electrical impulses transmitted through the radio to small motors or "servos" in the airplane. The servos physically move the rudder, elevator or other control in response to those commands.

What is the most important step in getting a model airplane ready to fly?

Probably the most important thing to do is to adjust the "trim settings" for the airplane. Models that aren't trimmed properly can be almost impossible to fly. RC transmitters normally come with trim controls to allow you to adjust the airplane's servos and control surfaces so that the plane will fly straight and level with the radio controls in the centered or neutral position and your hands off the radio.

What is "FPV flying"?

FPV stands for "First Person View" and refers to flying a model aircraft as if you were on board through the use of a miniature video camera and RC relays in the aircraft.

What is an "ARF" RC model airplane?

ARF stands for "Almost Ready to Fly" - ARF models generally require just a few additional items to be flight ready.

What rules apply to RC aircraft in the U.S.?

Federal Aviation Administration (FAA) Advisory Circular 91-57 states that radio-controlled model aircraft must stay below 400 feet altitude and that if you're flying your model with 3 miles of an airport you need to notIfy the airport. Considering the small size, weight and limited capabilities of most RC models, current FAA regulations are fairly relaxed. However, that may change due to the rise of RC models that are basically drones.

Where can I find RC drone kits?

- Http://www.nitroplanes.com/4eldronerqprra.html - 4-channel electric drone RQ/MQ-1 Predator 63 inch radio remote controlled RC spy plane ARF ($62).
- http://www.rchelicopterselect.com/drone-rq-1-predator-4-channel-electric-rc-airplane-kit.html – RQ/MQ-1 Predator 4-channel electric RC airplane kit ($92.99).
- http://nitroplanes.com/mq9.html – PROJET MQ-9 Reaper 98 inch 4-channel brushless remote control RC drone ARF kit V2 ($139.95). Additional options include a choice of FPV camera, charger and flight simulator.

What rule changes may be in store for RC aircraft?

According to recent (Feb. 2011) reports, the FAA may be about to impose much stricter regulations regarding the flying of RC model aircraft. Spurred by the growing number of sUAS (small Unmanned Aircraft Systems), new regulations may include a 55 pound weight limit, a ban on gas turbine engines, and an expanding of the "no fly without permission" rule on flying near airports, increasing the range from a 3 mile radius to 5 miles or more.

What about RC flight simulators?

Pilots generally start out by taking flying lessons; if you can find an experienced RC pilot to teach you that may be the way to go, but failing that, RC flight simulators can help you "earn your wings". A good quality RC simulator will consist of some very powerful software and may also include a realistic transmitter interface controller. Popular flight simulators like RealFlight and Phoenix feature a variety of aircraft to choose from, including helicopters, prop planes, fighter jets, airliners and gliders. Some simulator software even includes a virtual flight school where real-world pilots teach you the basics, from take-off and landing to advanced aerobatic maneuvers.

Is there a good RC flight simulator for $100 or less?

There are several. One of the best is RealFlight Basic which sells for right around $100. It comes with dozens of aircraft models to choose from, photo-realistic airfields, and a transmitter that has all the proper controls and functions and plugs into a USB port on your computer. Another is the *Phoenix R/C Pro Simulator V5.5 with DX6i Transmitter* by Runtime Games for around $220.

What are some other RC simulators?

- Manuel's Radio Control Flight Simulator – A free simulator that focuses on the basic

principles of flight control rather than fancy graphics
- ClearView RC Flight Simulator – A commercial simulator with a demo version (limited to two airplanes and a helicopter) that lets you try out the software. Buying the full version adds a host of other aircraft and even more user-submitted models can be downloaded from the website. You also have the option to fly in multi-player mode over the internet.
- RC Plane Master – This flight sim focuses on airplanes. The package includes a controller that plugs into a USB port and 12 aircraft, including electric, glow-plug and jet-powered models.
- E-Sky 4 Channel Flight Simulator Training Kit – The kit contains the software on a CD along with an E-Sky transmitter that plugs into your PC's USB port. E-Sky Flight Simulator is a low-cost way to get an introduction to RC flying.

What are some websites for RC aircraft?
- Http://www.rc-airplane-world.com
- http://rcvehicles.about.com
- www.nitroplanes.com
- www.rc-airplanes.com
- www.rcplanetalk.com (forum)
- www.hobby-lobby.com/radio_control_airplanes_196_ctg.htm

When does a radio-controlled model become a drone?

Technically, a model aircraft becomes a drone when it goes out of sight of the operator.

Are there any online drone simulators?
http://www.airforce.com/interactive/drone/index.html offers a very basic simulation that provides a rough idea of what it's like to fly a drone.

Drone Racing

What is FPV drone racing?
Think of a smaller, unmanned version of the pod racing in Star Wars. In First-Person View (FPV) racing, competitors maneuver small, durable drones equipped with onboard cameras around an obstacle course. Racers guide their drone by using a headset or console that receives the drone's video feed, so the racer sees exactly what the drone's camera sees. The "pilot" steers the drone through the course as quickly as possible and the competitor with the fastest time wins.

When and where did drone racing get its start?
Organized drone racing got its start as an amateur sport in Australia in 2014.

How fast do racing drones go?
Current models can reach speeds of up to 120 mph.

Are there particular types of drones used in drone racing?
There aren't any specific types, although the majority of competitors fly quadcopters.

How much does it cost to get into drone racing?
The low end is somewhere around $200 – that will buy you a ready-to-fly drone, which is all you need to get started. However, to have a chance to actually win races you will probably need to invest several times that much.

Are there different drone racing leagues?
There are a dozen or more well-known racing leagues – the major ones include:
- Drone Racing League (DRL) – Uses indoor, single-lap courses with LED-illuminated shapes as obstacles.
- DR1 Racing – Uses outdoor courses (usually in familiar locations around the world), with a combination of natural and man-made obstacles.
- MultiGP provides standards for local chapters to follow in setting up courses and also ranks MultiGP pilots worldwide on standard measured courses.

Which is the most popular racing league?
DR1 is currently the leader, with its events airing on broadcast TV in over 100 countries on over 300 media outlets including CBS, Fox Sports, the Discovery Channel, and Eurosport.

How much money can you make racing drones?
The inaugural World Drone Prix was held in Dubai in March of last year. The winner, Luke Bannister, a 15-year-old drone racing pilot from the UK, took home $250,000 (out of a total prize package of $1 million).

Drone Organizations

Are there any national or international drone organizations?
Here are a few:
- Association for Unmanned Vehicle Systems International (www.auvsi.org) – AUVSI is the world's largest non-profit organization (over 7,500 members in 55 countries) promoting the use of drones. It issues a monthly magazine "Unmanned Systems" to members, along with an email newsletter, the "Daily Robot". AUVSI also produces webinars on UAV issues.
- Unmanned Aerial Vehicle Systems Association (UK) (www.drones.org) – The oldest non-profit trade association devoted to the unmanned aerial systems community.
- Association of Commercial Unmanned Aircraft Systems (www.acuas.org) – ACUAS works to promote the commercial usage of small UAS (drones under 55 pounds). Its focus is on improving regulations and public communication in order to provide safe and beneficial sUAS commercial services.
- The Association of Professional Drone Pilots (www.prodronepilots.org) – Offers assistance to any public or private organization that wants to develop a drone program, to help them run their operation safely and legally.
- U.S. Association of Unmanned Aerial Videographers (www.uavus.org) – A professional organization for anyone involved in UAV imagery. UAVUS currently has over

6,000 members working in the fields of "aerial cinematography, precision agriculture, infrastructure inspection, real estate marketing, search and rescue, and more."

- Professional Society of Drone Journalists (www.dronejournalism.org) – Established in 2011, the PSDJ "is dedicated to establishing the ethical, educational and technological framework for the emerging field of drone journalism. We develop small unmanned aerial systems (sUAS) for journalists, and explore best practices to deploy them for a variety of reporting needs, including investigative, disaster, weather, sports, and environmental journalism."
- National Business Aviation Association/UAVs (https://www.nbaa.org/ops/uas/)

Drone Websites

AUVSI (Association for Unmanned Vehicle Systems International):
AUVSI (http://www.auvsi.org) is dedicated to promoting unmanned systems and related technologies. AUVSI is a non-profit organization and includes over 6,000 members representing 2200 companies in 55 countries.

sUAS News:
http://www.suasnews.com is a primary news source for information on smaller drones.

AOL drone News:
http://www.aolnews.com/tag/drone/ offers news, photos and background information on drones.

Military Industry News on drones:
http://military.einnews.com/news/drones provides news on developments in the drone industry.

UAV Coach:
http://community.uavcoach.com offers in-depth articles and online preparatory courses for drone operation. You will also find a number of forums including: drone pilot talk, news and events, regulations discussion and business advice.

RC Groups:
www.rcgroups.com/forums/index.php hosts forums on topics related to radio-controlled model aircraft

and a large portion of the forum discussions deal with drones.

Drone Pilots Forum:

http://dronepilots.zone operates several online communities for drone enthusiasts (membership is free).

Drone Forum:

http://www.rcgroups.com/drone-unmanned-aerial-vehicles-238/ offers forum posts regarding radio-controlled recreational drones and drones in general.

DIY Drones:

DIY Drones (www.diydrones.com) helps drone hobbyists construct their own unmanned flying vehicles. There are forums, blogs, helpful articles and an online store where you can buy complete kits or just the parts to build your own drone. DIY Drones offers a wide range of drone kits like the "BlimpDuino" (priced around $90), a flying balloon with an Arduino board that controls the balloon's propulsion and navigation. Other drones for sale include airplanes, helicopters, quadcopters and blimps.

Games/Videos/Books

Are there drone video games?
There are a few drone games out there. For example:
- drone Predator ($28.53 at amazon.com) (Windows Vista/2000/XP) – Fly a Predator from a ground station.
- drone Fighter (iPhone) (Apple App Store)
- The web site http://www.airforce.com/interactive/drone/index.html lets you fly a simulated drone mission.

Any books featuring drones?
There are a number of non-fiction books about drones and Dale Brown has authored several fiction books in the "Dale Brown's Dreamland" series that feature futuristic drones called "Flighthawks".

Can playing video games help me fly drones?
Researchers at the Massachusetts Institute of Technology (MIT) and other groups are trying to determine if there's a link between video game expertise and superior drone controllers. So far no direct correlation has been found, but Senior Chief Aviation Electronics Technician Stephen Diets who is training to be an operator of the MG-8 Fire Scout feels that his background playing games on the Atari and Playstation is helpful. According to Diets, the things you have to keep track of in some video games (maps, compasses, ammo loads) are the same type of things you have to deal with in controlling a military drone. Games like "Tom

Clancy's Ghost Recon – Advanced Warfighter" even let players fly a drone over a digital battlefield to detect threats. Just be aware – actually flying a military drone isn't nearly as exciting as a video game. It's serious business and requires the ability to concentrate and stay alert over a long period of time.

Is the U.S. Military making use of Video game technology for drone training?

Realistic video game simulations are rapidly becoming a part of military drone training courses. For example the Army's drone training courses at Fort Huachuca now let trainees spend a considerable amount of time flying simulated drones on virtual missions over areas similar to that in Iraq and Afghanistan. Each simulator uses digitized satellite imagery from actual locations in Iraq and Afganistan to provide realistic terrain and instructors can input simulated targets or items of interest. Trainees do mission planning based on intelligence provided to them, make out a list of "targets of opportunity" to check out. Then they "fly" the mission, attempting to find the targets of opportunity (not necessarily targets to be fired on) and anything else pertinent. Once the mission is over, instructors review the results and critique the trainees on their performance.

Also, a company named L-3 Link has created Predator and Reaper simulations (used by the Air Force and Air National Guard) that mimic different times of day, various weather conditions, and include life-like characters and realistic scenarios. Ultimately the military would like to have simulations that trainees could take home with them, run the games on their own PC and study various situations and outcomes – a method of providing realistic "homework" in other words. The center also trains mechanical and electronic technicians to provide maintenance for the drones.

Appendix A - Terminology

2.4 Ghz (gigahertz) Spread Spectrum – The number of gigahertz associated with a drone controller describes the radio frequency rate used by the controller and the 2.4 gigahertz spread spectrum controllers have become a standard for low budget drones. Because the spread spectrum is "frequency agile," it can "hop" to different frequencies to account for any atmospheric or other issues which means fewer crashes.

250 Racer: A term used to describe a racing multicopter that's 250mm diagonally from end to end. 250 Racers are a very common type of quadcopter used by FPV (First Person View) drone racers.

808 Camera – A common name for the small 'spy' cameras frequently used on drones for taking video.

Accelerometer - A device that measures the acceleration forces in a certain direction. Accelerometers are used to help stabilize a drone and help maintain the drone's orientation.

AGL – Altitude above Ground Level.

AHRS – Altitude Heading Reference System

ALR – Automatic Launch and Recovery

ATARS - Advanced Tactical Airborne Reconnaissance System

ACC – Air Combat Command

AFCS – Automatic Flight Control System

AGL – Above Ground Level

ASIP – Airborne Signals Intelligence Payload

ATO – Air Tasking Order

AUVSI – Association for Unmanned Vehicle Systems International.

Barometric Pressure Sensor – A device that uses barometric pressure readings to determine the altitude of the drone; allows to drone to stay at or below a given altitude.

Bind – The process of getting the controller (transmitter) to communicate with the drone.

BLOS – Beyond Line Of Sight (communications).

BVLOS – Beyond Visual Line of Sight.

CAS – Close Air Support.

DARPA – Defense Advanced Research Projects Agency.

ESC (Electronic Speed Controller) – A component of a drone that translates requests from the flight controller into speed adjustments on the rotors.

(Sometimes referred to as an EMC – Electronic Motor Controller).

EW – Electronic Warfare.

FAA – Federal Aviation Administration.

FAR – Federal Aviation Regulation.

FCS – Flight Control System.

FL – Flight Level (refers to altitudes above 18,000 ft)

FLIR – Forward Looking Infrared.

Fly-Away Protection System – An on-board system that will either land the drone safely or keep it within a predetermined area if the link between the operator and the drone is lost.

FPV – First Person View (also, Remote Person View). Refers to the use of a camera mounted on a drone which sends video back to the operator, allowing him or her to see exactly what the drone sees.

GCS – Ground Control Station. A GCS normally controls a UAV during flight. It transmits in-flight commands to the UAV and receives sensor and video data from it.

Geofencing - The use of GPS technology to create a virtual geographic boundary, enabling software to trigger a response when a drone enters or flies within a particular area.

GIS – Geographic Information System.

GPS – Global Positioning System.

Gyroscope – A device that measures the rate of rotation of the drone and helps to maintain the orientation of the drone and keep it balanced correctly with respect to yaw, pitch and roll.

HALE – High Altitude Long Endurance. A type of UAV designed for high altitude, long range missions.

Head tracking - A feature on some drone controller goggles that allows you to maneuver your drone's camera angle during flight by moving your head up and down or side to side.

IFR – Instrument Flight Rules.

IMU – An Inertial Measurement Unit is a combination gyroscope and accelerometer which helps a controller to orient and stabilize the drone it's controlling.

INS – Inertial Navigation System. An electronic system that utilizes a combination of accelerometers and gyroscopes (and occasionally magnetometers) to continuously calculate the position, orientation and velocity of an aircraft.

ISR – Intelligence, Surveillance and Reconnaissance

JTAC – Joint Terminal Attack Controller

LALE – Low Altitude Long Endurance

LOS – Line Of Sight (communications).

LRE – Launch and Recovery Element.

mAh (milliamp hours) – A unit of measurement that indicates how much power a battery can provide before it has to be recharged.

MALE – Medium Altitude Long Endurance.

MARS – Mid-Air Retrieval System.

Micro – A general term referring to small drones, usually weighing around 25 grams.

NAS – National Air Space

NOTAM – Notices to Airmen

MCE – Mission Control Element

OIF – Operation Iraqi Freedom

OEF – Operation Enduring Freedom (Afghanistan).

Part 107 - Refers to Part 107, Title 14 of the Code of Federal Regulations (14 CFR) for non-hobbyist unmanned aircraft operations in the U.S. Part 107 allows for routine operation of small Unmanned Aircraft Systems (sUAS) in the National Air Space (NAS) and provides safety regulations for those operations.

PDB (Power Distribution Board) – Drone component that distributes battery power to the different parts of the drone.

PID – Proportional, Integral, Derivative control. PID refers to the mathematical process used by a flight controller to achieve a stable power to response ratio in drone motors. Adjusting PIDS can make a drone more responsive, but also can make it less stable.

Pitch – The degree to which a drone's nose is up or down relative to an axis running from wingtip to wingtip.

Quadcopter or "Quad" - A drone with four motors and propellers which are arranged in a square to provide smoother, more stable flight.

RATO – Rocket Assisted Take Off

ROA – Remotely Operated Aircraft

RPAS – Remotely Piloted Aerial System

RPAV – Remotely Piloted Aerial Vehicle

RPV – Remotely Piloted Vehicle

R/C (or RC) – Radio Controlled

SLAR – Side-Looking Airborne Radar

STUAS – Small Tactical Unmanned Aircraft System

sUAS – Small Unmanned Aircraft System

UAS – Unmanned Aerial System

drone – Unmanned Aerial Vehicle

UCAV – Unmanned Combat Air Vehicle

VFR – Visual Flight Rules

VLOS – Visual Line Of Sight

WAAS – Wide Area Augmentation System

Appendix B – Types of Drones

Note: The military naming scheme for drones works as follows: the letter "R" stands for Reconnaissance, the letter "M" stands for Multi-role, "Q" stands for an unmanned aircraft system, and the addition of a letter "A", "B", etc. at the end of the name indicates a modified or enhanced version of the original drone.

MILITARY:
- **RQ-11B Raven** - The Army has more than 7,000 hand-launched Raven drones in service, making it the most prevalent drone around (and the Raven is now in use by the Marine Corps as well). The Raven is 3 feet long and weighs 4.2 pounds and is typically fitted with an electronically stabilized color video camera or an infrared video camera for night missions, providing ground troops with an overview of their immediate surroundings.. Effective range for the RQ-11B Is about six miles.
- **Wasp III** - Used by U.S. Special Forces units, the Wasp III is electric-powered and extremely quiet. It is hand-launched, weighs just one pound, flies 20 to 40 miles per hour at altitudes up to 500 feet and can be programmed to fly a completely autonomous mission.
- **Desert Hawk** - Used by American and British troops in Afghanistan, the Desert Hawk is hand-launched and follows pre-programmed coordinates to give troops an "over-the-hill" view day or night up to six miles away.

Extremely durable, the Desert Hawk is built of expanded polypropylene and fitted with Kevlar skids.

- **T-Hawk** - Used by U.S. Army infantry in Iraq. The VTOL T-Hawk can fly at altitudes as high as 10,000 feet for up to 45 minutes.
- **Scan Eagle** - Used by U.S. Marines, the Scan Eagle is 4 feet long and weighs about 40 pounds. It's powered by a gasoline engine and launches from a catapult.
- **RQ-7 Shadow** - Used in Irag and Afghanistan to provide tactical surveillance. The Shadow launches from a catapult and can stay aloft for five to six hours, flying at altitudes up to 14,000 feet. It's about 11 feet long, weighs around 375 pounds and is equipped with an infrared illuminator, allowing it to provide laser targeting for laser-guided missiles and bombs. Note: The Shadow can also deliver a 20 lb. "Quick-Meds" canister to front-line troops.
- **MQ-1 Predator** - Used extensively by the United States in Iraq, Afghanistan and Pakistan. The Predator is an armed reconnaissance drone which can carry two Hellfire air-to-ground missiles. It has a range of about 400 nautical miles and can stay airborne for up to 24 hours.
- **FINDER** (Flight Inserted Detection Expendable for Reconnaissance) - The FINDER is actually a drone which can be launched from another drone (a Predator), to get a closer view of a target. It is just over 5 feet in length, weighs about 60 pounds and can be flown using the

Predator's controls. The FINDER launches like a rocket and then its wings unfold for flight.

- **MQ-1C Gray Eagle** - An upgrade to the MQ-1 Predator developed for the U.S. Army, the Gray Eagle is an Extended Range Multi-Purpose (ERMP) drone powered by a diesel engine that burns jet fuel. It is capable of operating for up to 36 hours at a time with a maximum altitude of 25,000 feet and an operating range of 200 nautical miles. The MQ-1C's nose houses a Synthetic Aperture Radar/Ground Moving Target Indicator (SAR-GMTI) system and an AN/AAS-52 Multi-spectral Targeting System (MTS). It can carry a payload of 800 pounds and can be armed with AGM-114 Hellfire missiles and GBU-44/B Viper Strike laser guided bombs.
- **MQ-5 Hunter** - Flown by the U.S. Army in Iraq and Afghanistan. The Hunter has been around for a while and was recently refitted as the MQ variant to carry Viper Strike munitions. It has a 34 foot wingspan, can fly 18 hours at altitudes up to 18,000 feet and can be flown from the same ground control station as the Army's version of the MQ-1 Predator.
- **MQ-8B Fire Scout** – The Fire Scout is an unmanned autonomous helicopter designed to provide reconnaissance and targeting support. The MQ-8B is fitted with a four-blade main rotor to reduce noise and improve lift. It can carry Hellfire missiles, Viper Strike laser-guided weapons, and the "Advanced Precision Kill Weapon System (APKWS), a laser-guided

70 millimeter folding-fin rocket. The MQ-8B is used by both the Army and the Navy.

- **MQ-9 Reaper** - The Reaper is a bigger, beefier cousin of the MQ-1 Predator and is also in use in Iraq, Afghanistan and Pakistan. Twice the size of a Predator, the Reaper is jet-powered and can carry 3,000 pounds of ordinance, including Hellfire missiles and the GBU-12 laser-guided bomb. After launch the Reaper is normally flown via satellite link by crews in the United States.

- **RQ-4 Global Hawk** - Used by U.S. forces in Iraq, Afghanistan and Pakistan, the Global Hawk is 44 feet long, weighs about 8500 pounds and can stay aloft up to 36 hours. It can fly at altitudes up to 65,000 feet and provide coverage of over 40,000 nautical square miles. The RQ-4 is powered by an Allison Rolls-Royce turbofan engine and carries a payload of 2,000 pounds. The Global Hawk features a full suite of infrared, electro-optical and radar sensors and can operate autonomously day or night in all types of weather.

- **RQ-170 Sentinel** - The highly-classified Sentinel flies sorties over Afghanistan (and possibly elsewhere), controlled by satellite link from Creech AFB in Nevada. It's an offspring of Lockheed Martin's Skunk Works program, but very little else is known about it. It is rumored to have been used in trying to track down Osama Bin Laden.

- Boeing ScanEagle – A small, long-endurance, low-altitude unmanned aerial vehicle (UAV) built by Insitu, a subsidiary of Boeing, used for reconnaissance

- **CU-170 Heron** – A Medium Altitude Long Endurance (MALE) drone developed by Israel, the Heron can stay airborne for over 50 hours at altitudes up to 35,000 feet. The Heron is capable of either pre-programmed flight or ground controlled flight and has fully automatic launch and recovery (ALR) and all-weather capabilities. It can carry a variety of sensors and can also be used for target acquisition and artillery adjustment.
- **Hermes 450/Watchkeeper** - Used by Israel and by British troops in Iraq and Afghanistan, Hermes can stay aloft for about 20 hours providing real-time surveillance to commanders in the field.
- MQ-25 Stingray – The Navy is developing the MQ-25 UCAAS (Unmanned Carrier Aviation Air System) to serve primarily as an airborne tanker. The Super Hornet sized Stingray will also be able to drop bombs and launch missiles, but aerial refueling will be its main mission.

EXPERIMENTAL:

- **Phantom Ray** - The Phantom Ray is a jet-powered stealth drone, shaped like a flying wing. It is currently being used to test advanced drone capabilities like radar jamming, aerial refueling, and air-to-air missile combat. The Phantom Ray has a ceiling of about 40,000 feet and an estimated maximum speed of a little over 600 miles per hour.

- **Demon** - A wing-shaped, highly maneuverable drone being developed by BAE Systems in England.
- **Vulture** - Vulture is Lockheed Martin's design for DARPA's Vulture program. It is designed to stay aloft for up to five years to provide continuous monitoring of some area for long periods of time. Vulture is equipped with a suite of day-and-night cameras that can scan a 600-mile area, sending imagery and other data back to handlers on the ground. A semi flexible structure allows the Vulture to bend instead of breaking when high winds cause it's lengthy frame to oscillate violently.
- **Global Observer** – AeroVironment is developing the Global Observer, a hydrogen-powered high-altitude long-endurance (HALE) drone. Global Observer will be capable of flights of at least 5 days in duration at altitudes up to 65,000 feet. Note: A GO-1 Global Observer crashed at Edwards AFB on April 1, 2011 while undergoing testing and future funding for the program could be a problem.

CIVILIAN (Commercial):
- **DJI Phantom** – The DJI Phantom series of camera-carrying drones has been a market leader since 2013. Current models are the Phantom 4 (priced at about $900) and the Phantom 4 Pro (priced at about $1500). Additional models include the DJI Inspire and the DJI Mavic Pro.
- **Yuneec Typhoon** – The Typhoon Q500, introduced in 2014, was the first ready-to-fly

out of the box drone. Current models include the Typhoon H 4k (a hexacopter with collision avoidance technology) and the Yaneec Breeze, a compact smart drone.

- **Parrot AR** – The Parrot AR (current model is the AR-2) is a mid-range hobby drone with an integrated FPV (First Person View) system. Other models include the Disco FPV, a fixed wing drone with 45 minute battery life and FPV goggles and the Mambo mini-drone.
- **Yamaha RMAX** – The RMAX is a 12-foot long unmanned helicopter and is in use by the city of Tucson, Arizona as a platform to spray wetlands for mosquito control.
- **SkySeer** – An electric-powered, short-range, hand-launched drone is being experimented with as a means of crime prevention. A complete SkySeer system retails for $25,000 to $30,000 including the drone, its electronics, video surveillance equipment, ground station and computer inferfaces.
- **Aerosonde** – A small drone designed to collect weather data (temperature, air pressure, wind and humidity) over oceans and remote areas, the Aerosonde was the first drone to cross the Atlantic. Launched from the roof rack of a moving car, the Aerosonde flew from Newfoundland, Canada to an island off the coast of Scotland in less than 27 hours, burning about 1 ½ gallons of gasoline. Aerosondes were also the first drones to fly into the eye of a tropical cyclone. The Aerosonde is about 5 ½ feet long with a wingspan of 9 ½ feet, a maximum speed of 90

miles per hour and a maximum altitude of 15,000 feet.

Appendix C – R/C Ground School

A good first step to learning to fly drones is to learn how to fly a radio-controlled (R/C) airplane. Here's a quick "ground school" on flying R/C models.

R/C airplanes are controlled by a radio signals transmitted from a hand-held transmitter to a receiver in the airplane. Older models used transmitters in the megahertz range but 2.4 gigahertz (GHz) radios are becoming standard since they are far less susceptible to interference.

Whatever type of transmitter is used the signals are picked up by the receiver and converted into physical movement through the use of servos inside the airplane that are linked to the model's control surfaces.

The control surfaces are used to change the plane's position or altitude or to correct for unintentional changes. When an airplane changes position in flight it rotates around one or more of three axes (imaginary lines that pass through the aircraft's center of gravity). Motion around the lateral axis (motion up and down) is referred to as **pitching**. Motion around the longitudinal axis is called **rolling**, and motion around the vertical axis (motion to one side or the other) is called **yawing**.

The primary control surfaces are the **rudder, elevator and ailerons**. The most basic R/C models (single channel models) will only control the rudder (the hinged portion of the vertical stabilizer on the tail of the airplane. The **rudder** controls the **yaw –** when the rudder moves to the left the plane turns left and vice versa.

The **elevators** are the hinged sections of the horizontal stabilizer on the tail of the airplane and they control the **pitch attitude** of the plane – whether the nose of the plane is pointing up, down or level. Moving the elevators upward pushes the airplane's tail downward and the nose up (increasing the angle between the plane and the streaming air – the **angle of attack**) and the plane starts to climb. Moving the elevators downward causes the nose to come down and the airplane will start to dive.

The **ailerons** are the hinged sections of the trailing edge of each wing. Ailerons always move in the opposite direction to each other – when the aileron on the left wing goes up, the aileron on the right wing goes down and vice versa. Ailerons control the **roll** of the airplane about its longitudinal axis. Moving the right aileron up decreases the lift on the right wing and causes the left aileron to move downward (increasing the lift on the left wing) and the plane rolls to the right. Moving the right aileron down (and the left aileron up) rolls the aircraft to the left. A side effect of using the ailerons is **adverse yaw** – a yawing movement in the opposite direction to the roll caused primarily by the change in **drag** on the left and right wings. Adverse yaw is compensated for by use of the **rudder**.

Some RC airplanes also have **flaps**, which are also hinged sections on the trailing edge of the wing. If the airplane happens to have flaps, they will be separate from the ailerons and located on the inside portion of the trailing edge of the wing. The ailerons will be on the outside portion of the trailing edge. Flaps work together – they go up or down at the same time and serve to increase or decrease the airflow over the wings, increasing or decreasing lift. Flaps are used primarily in landing; lowering the flaps increases the lift under the wing and helps keep the airplane in the air as it slows down to land.

Because of their design some aircraft don't have all of these separate control surfaces – they use control surface **mixing**. For example, the elevators and ailerons may be combined or "mixed" and are called **elevons**. Elevons look like elevators but they can move together (like elevators) or separately (like ailerons). Similarly **flaperons** mix the actions of flaps with ailerons, with one pair of control surfaces along the trailing edge of the wing doing the job of both aileron control and flap control.

Every controllable operation on an R/C plane is referred to as a **channel**. A single channel model can only control one type of action – motor (on or off), rudder, ailerons, etc. A 4 channel model might be able to control the rudder, elevator, ailerons and the plane's motor.

However many channels your R/C plane has, the first step in flying your plane is to go through a pre-flight checklist:
- Make sure all screws or other fasteners are tight

- Check to make sure all control surfaces (rudder, ailerons, etc.) move freely and respond properly to commands from the transmitter
- make sure the receiver, battery pack or other electronic gear is fastened securely inside the plane
- If the plane is powered by an internal combustion engine check the fuel tank and tubing – if the plane is battery powered make sure the batteries are fully charged
- Check to make sure no wires or other items are in position to interfere with the servos
- Make sure the receiver is receiving commands from the transmitter

Once you've gone through the pre-flight checklist it's time to get your plane airborne.

To launch the plane, make sure it's facing directly into the wind. Always take off into the wind because that maximizes the airflow over the wings and helps the aircraft get airborne sooner. Slowly increase power to full, letting the airplane accelerate over the ground. You will probably need to use the rudder to keep the plane moving in a straight line – don't let it veer off to the left or right.

The plane should take off fairly quickly – if it's struggling to get airborne, try giving it a small amount of up elevator to lift the nose. Don't apply full elevator as the plane starts its takeoff run – the plane will likely take off, climb at too steep an angle, stall out and crash. Once the plane is airborne remember to execute a turn before it goes out of range of the transmitter.

Hand launching works basically the same as letting the plane take off on its own. Hold the airplane about head-high and bring the motor to maximum power, then take a step forward and give the plane a firm push, keeping the aircraft level or very slightly nose-up.

Once the model is airborne you need to **"trim"** the airplane. Use the trim controls on the transmitter to correct any tendency of the airplane to turn, bank, climb or dive on its own.

To land the plane, be sure to land by flying a **"downwind leg"** (with the wind) and then turning into the wind. Landing into the wind increases the airflow over the wings and helps keep the plane from stalling out as it slows down. As you turn into the wind start to slow the plane down and use the rudder to keep it in a straight line. As the airplane nears the ground reduce power to the motor completely and apply a little up elevator to slow the plane's speed and rate of descent further, until it touches down. This maneuver is called **"flaring"** and the timing is critical – flare too soon and the plane will stall and crash, too late and you're going to have a very rough landing.

Basic maneuvers include:
- Roll – to execute a roll start by flying straight and level on at least half throttle. Apply a small amount of up elevator and left or right

aileron (depending on which direction you want to roll). As the plane starts to roll inverted, stop applying up elevator, then once the plane is fully inverted, use a very small amount of down elevator to maintain altitude and let the airplane complete the roll.

- Inside Loop – Start by flying straight and level into the wind, then open the throttle and apply full power while pulling back gently on the elevator stick to start a climb. Let the plane climb until it starts to roll over onto its back, then close the throttle and keep holding the elevator stick back. As the plane completes the loop move the elevator stick back to neutral to resume level flight.
- Outside Loop – Start the loop by rolling the plane inverted (upside down) and using down elevator to begin climbing. Reduce power at the top of the loop and as the plane completes the loop roll the airplane right side up again.

Appendix D – Flight Instruments

Most aircraft have certain standard flight instruments (although in the case of drones these instruments will appear as readouts on a workstation screen rather than as actual gauges):

- **Altimeter** – Shows the aircraft's altitude above sea-level by measuring the pressure difference between a stack of aneroid capsules inside the altimeter and the atmospheric pressure outside the aircraft.
- **Airspeed Indicator** – Shows the aircraft's speed (usually in knots) relative to the surrounding air. The airspeed indicator has to be corrected for the density of the air, which varies with temperature, altitude and humidity, to get the true airspeed and for wind conditions to get the speed over the ground.
- **Attitude Indicator (Artificial Horizon)** – Shows the aircraft's attitude (position in space) relative to the horizon. The attitude indicator lets the pilot know if the wings are level and if the aircraft's nose is pointing above or below the horizon. This instrument is particularly important in bad visibility or for instrument flight.
- **Magnetic Compass** – Shows the aircraft's heading relative to magnetic north. Because turning, climbing, descending or accelerating can cause the magnetic compass to give confusing indications, the **Heading Indicator**

is also used to determine the aircraft's true heading.

- **Heading Indicator (Directional Gyro or Gyrocompass)** – Shows the aircraft's heading relative to geographical north. The Heading Indicator is basically a spinning gyroscope and as such is subject to drift errors (precession) which are normally corrected periodically by calibrating the heading indicator to the magnetic compass. In more advanced aircraft the heading indicator has been replaced by a **Horizontal Situation Indicator (HSI)**.
- **Turn Indicator** – Shows the direction of turn and the rate of turn.
- **Vertical Speed Indicator** – The VSI displays changing air pressure as a rate of climb or descent.

In addition to these instruments the controls in a UAS will also include items such as autopilot status, GPS display, data link readouts, digital maps and waypoint displays, telemetry data and video images.

Appendix E - Remote Pilot Study Guide

Passing the FAA examination for a Remote Pilot Certificate allows you to act as the Pilot In Charge (PIC) for commercial drone operations. The exam itself covers 13 basic areas:

- **Regulations**
  - 14 CFR Part 107 definitions
- **Airspace Classification, Operating Requirements and Flight Restrictions**
  - Controlled airspace
  - Uncontrolled airspace
  - Special use airspace
  - Other airspace areas
  - Air Traffic Control and the National Airspace System (NAS)
  - Visual Flight Rules (VFR) and symbols
  - Notices to Airmen (NOTAMS)
- **Aviation Weather Sources**
  - Surface aviation weather sources
  - Aviation weather reports
  - Aviation forecasts
  - Convective Significant Meteorological Information (WST)
- **Effects of Weather on sUAS Performance**
  - Density altitude
  - Performance
  - Measurement of atmosphere pressure
  - Effect of obstructions on wind
  - Low-level wind shear
  - Atmospheric stability
  - Temperature/dew point relationship

- Clouds
- Fronts
- Mountain flying
- Structural icing
- Thunderstorm life cycle
- Ceiling
- Visibility
- **Small Unmanned Aircraft Loading**
  - Weight
  - Stability
  - Load factors
  - Weight and balance
- **Emergency Procedures**
  - Inflight emergencies
- **Crew Resource Management**
- **Radio Communications Procedures**
  - Understanding proper radio procedures
  - Traffic advisory practices at airports without operating control towers
- **Determining the Performance of Small Unmanned Aircraft**
  - Effect of temperature on density
  - Effect of humidity (moisture) on density
- **Physiological Factors Affecting Pilot Performance**
  - Physiological/medical factors that affect pilot performance
  - Vision and flight
- **Aeronautical Decision-Making and Judgment**
  - History of aeronautical decision making (ADM)
  - Risk management

- Crew resource management and single-pilot resource management
- Hazard and risk
- Human factors
- The decision-making process
- Decision-making in a dynamic environment
- Situational awareness
- **Airport Operations**
  - Types of airports
  - Sources for airport data
  - Latitude and longitude (meridians and parallels)
  - Antenna towers
- **Maintenance and Preflight Inspection Procedures**

Here's a rough breakdown of the exam by topic:
- Regulations: 15% - 25%
- Airspace & requirements: 15% - 25%
- Weather: 11% - 16%
- Loading & performance: 7% - 11%
- Operations: 35% - 45%

Items to remember about the exam:
- You must be at least 16 years old to take the exam.
- It costs $150 to take the exam and the certification is good for 2 years.
- You need to answer 70% of the questions correctly in order to pass.
- The exam consists of 60 multiple-choice questions. (Note: there may be more than 60 questions if the FAA is evaluating a few new

ones – however you still only need to get 42 of them right).

- To schedule to take the exam go to: https://www.faa.gov/training_testing/testing/media/test_centers.pdf for a list of examination centers. Pick one near you and call one of the 800 numbers listed on the web page in order to set a time to take the test.
- There are numerous websites that offer paid training to help you prepare for the certification exam – just search for "remote pilot certification training". There are also many free resources you can use to help you study for the exam including:
  - The FAA's Remote Pilot Study Guide (https://www.faa.gov/regulations_policies/handbooks_manuals/aviation/media/remote_pilot_study_guide.pdf)
  - Remote Pilot Test Prep 2018: Study & Prepare: Pass your test and know what is essential to safely operate an unmanned aircraft – from the most trusted source in aviation training (Test Prep Series) by ASA Test Prep Board
  - https://www.youtube.com/watch?v=6_ucCKFJUCU – Tony & Chelsea Northrup's FREE Drone Certification Study Guide: FAA Part 107 sUAS Test

Appendix F – Buying a Drone

Here are a few of the features to look for, depending on what you want in a drone:

- Ready-to-fly versus Bind-and-fly. Ready-to-fly drones are just that – ready to go (although you may have to attach the batteries or propellers). Bind-and-fly drones are generally for people who already have a controller or control system of their own and want to connect ("bind") the drone to their controller before starting to fly.
- Cost. Drones under $180 are generally considered "inexpensive" models. Even though they're lightweight battery life for these drones is normally no more than 10 minutes (which means you may want to have multiple batteries available). Drones in this class also don't have 3-axis stabilization so you won't have stabilized images and video. Popular drones in the less-than-$180 category include the Hubsan X4, the Parrot Rolling Spider, the U49W Blue Heron and the Blade Nano. If you're searching for something with more professional-level features, you're probably looking at drones that run from $200 to several thousand dollars, including the DJI Mavic Pro, the DJI Spark, the DJI Inspire, the DJI Phantom (notice a trend here?), and the Yuneec H520.
- Configuration. For straight-line speed a fixed-wing drone is probably the way to go. For most other uses a quadcopter on a X or H

frame offers the most stable platform and is probably the best choice, especially for beginners.

Alphabetical Index

Notes: